普通高等教育机电类教材

实用机械工程专业英语

（第2版）

杨春杰　蔺绍江　何　彬　主　编

修　云　王培玲　吴小艳　副主编

华　江　云丰泽　柯　奇　参　编

電子工業出版社

Publishing House of Electronics Industry

北京 · BEIJING

内 容 简 介

本书内容主要以机械工程专业领域的基础理论和实际应用的英文文章为主，包括工程图学及计算机辅助设计、机械设计、机电技术、汽车工程、材料成型、先进制造技术、科技论文的写作和设备说明书翻译实例，可满足机械工程类不同专业如机械设计与制造、机械设计制造及其自动化、交通运输、数控技术、材料成型及控制工程、机械电子等专业的专业英语课程的教学、自学和参考用。

本书的内容选材篇幅有长有短，难易适中，涵盖面广。课后附有生词表和难句分析，可适应专科生、本科生、硕士研究生等不同层次的需求。

图书在版编目（CIP）数据

实用机械工程专业英语 / 杨春杰等主编. —2 版. —北京：电子工业出版社，2019.6
普通高等教育机电类“十三五”规划教材
ISBN 978-7-121-36163-0

Ⅰ. ①实… Ⅱ. ①杨… Ⅲ. ①机械工程—英语—高等学校—教材 Ⅳ. ①TH

中国版本图书馆 CIP 数据核字（2019）第 051818 号

责任编辑：郭穗娟
印　　刷：涿州市般润文化传播有限公司
装　　订：涿州市般润文化传播有限公司
出版发行：电子工业出版社
　　　　　北京市海淀区万寿路 173 信箱　邮编　100036
开　　本：787×1092　1/16　印张：15　字数：476 千字
版　　次：2013 年 8 月第 1 版
　　　　　2019 年 6 月第 2 版
印　　次：2023 年 7 月第 6 次印刷
定　　价：59.80 元

凡所购买电子工业出版社图书有缺损问题，请向购买书店调换。若书店售缺，请与本社发行部联系，联系及邮购电话：（010）88254888，88258888。
质量投诉请发邮件至 zlts@phei.com.cn，盗版侵权举报请发邮件至 dbqq@phei.com.cn。
本书咨询联系方式：（010）88254502，guosj@phei.com.cn。

前　言

本书共 7 个单元，每个单元包含 2～7 课，每课可供 2～3 个学时讲解，各专业可根据不同的需要选取教学内容。本书内容主要以机械工程专业领域的基础理论和实际应用的英文文章为主，包括工程图学及计算机辅助设计、机械设计、机电技术、汽车工程、材料成型、先进制造技术、科技论文的写作和设备说明书翻译实例，可满足机械工程类不同专业的专业英语的教学、自学和参考用。本书内容涵盖的专业有机械设计、机械制造、汽车构造与维修、交通运输、模具设计与制造、材料成型与控制、机械电子、先进制造技术、机械加工设备等，内容由简至深，既有基础理论也有实践应用，也涉及最先进的科技前沿动态和创新思想。本书侧重应用型人才培养所需的行业领域专业英语的学习和应用。

本书是在编者近 20 年的专业英语教学经验总结和素材积累的基础上编写而成的，力求满足专科生、本科生、研究生等不同层次的教学和自学需求。选材注重机械工程专业英语的实用性，结合大量的实例展开教学内容，易于教学和自学。每篇课文后配有习题，可考查学生对内容的理解和掌握情况。本书还配有参考译文（可登录华信教育资源网：http://www.hxedu.com.cn，免费下载译文），译文经过反复修改与斟酌，较好地表达了原文的内容。但多年的教学经验告诉我们，教师或读者在学习原文时，尽量不看参考译文，从原文获取原汁原味的信息最好。

本书由杨春杰、蔺绍江、何彬担任主编，修云、王培玲、吴小艳担任副主编，华江、云丰泽、柯奇参编。尽管编者为编写本书付出了心血，但由于时间有限，无法做到尽善尽美。书中如有疏漏和不当之处，敬请读者批评指正。编者在此表示衷心的感谢，并将继续努力，不断地修订和完善本书。

编　者

2019 年 3 月

目　录

Unit One　Engineering Graphics and CAD

Lesson 1　Engineering Graphics in the Third-angle Projection

教学目的和要求

通过本文的学习，了解第三角画法的概念、投影原理和应用。要求掌握工程图学的各种视图及投影规律的英文专业术语和表达习惯，了解第三角投影和第一角投影画法的区别和变换。

重点和难点

（1）了解第三角投影的原理和投影规律。
（2）掌握第三角画法的组合体、零件图、装配图的专业词汇和表述。
（3）难点在于第三角投影和第一角投影画法的变换。

Part A　Text

Preface

Graphics comes to our vocabulary from the Greek word grapho, whose extended meaning is *drafting* or *drawing*, the drawing is the primary medium for developing and communicating technical ideas[1]. Engineering drawings provide an exact and complete description of objects. In addition to a description of the shape of an object, an engineering drawing gives all further information needed to manufacture the object drawn, such as dimensions, tolerances, and so on. So, it is often said the engineering drawing is the common language of engineering. Every engineer must master this language. The main tasks of engineering graphics: to learn the knowledge of the projections to cultivate the drawing making and drawing reading abilities to cultivate and develop the spatial analysis and spatial visualization abilities.

As a common language of engineering，the drawing is used to direct the production and make the technical interchange. So, it is necessary to specify, in a unified way, drafting practices, such as the layout of drawings, dimensioning, and so on.

★ **Formation of three-projection views**

Three projection planes system: Since it is impossible to determine the position of a point with its only one projection, more projection planes are added. Usually, three projection planes perpendicular to each other are used in orthographic projection. They are horizontal projection plane, frontal projection plane and profile projection plane, denoted by H, V, W, respectively.

The third-angle projection: Three projection planes divide space into eight parts or quadrants numbered from 1 to 8, as shown in Fig.1.1. According to the Chinese National Standard of Technical Drawings, the first-angle projection is used to make engineering drawings while in some other countries, such as in the USA and Canada, the third-angle projection is used[2]. In this paper we focus on the third-angle projection.

Formation of three-projection views: In first angle, an object is placed in quadrant 1, and observer always looks through the object towards the projection plane. But in third-angle, the object is placed in quadrant 3, and observer always looks through the projection plane towards the object. In third-angle projection, projection plane is assumed transparent, so form into views.

The symbols of the first -angle and the third -angle projection in the following Fig.1.2

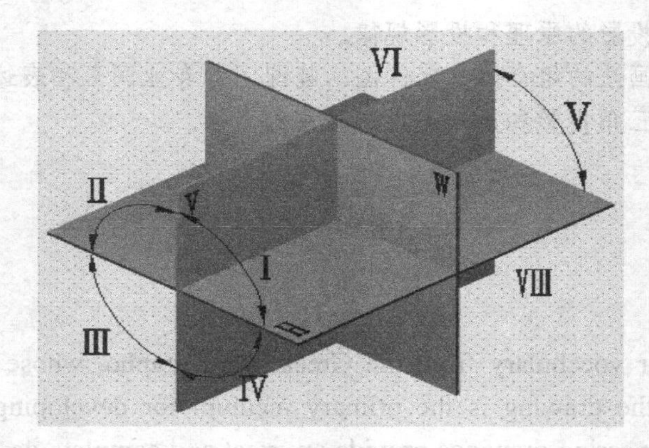

Fig.1.1 Projection planes

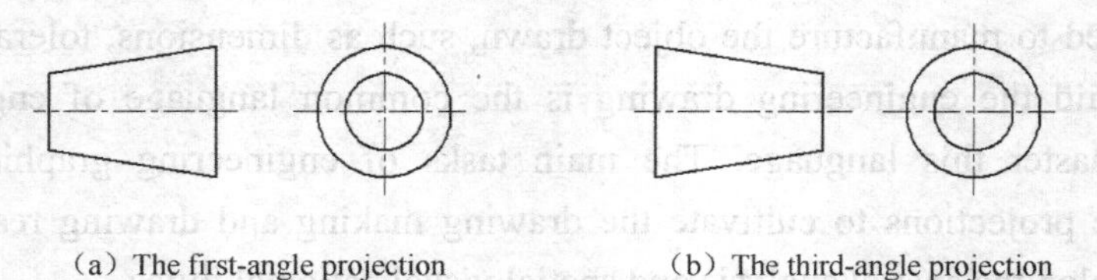

(a) The first-angle projection (b) The third-angle projection

Fig.1.2 The symbols of the different angle projection

★ **Composite Objects**

Projection rules of an object: The front and top views are aligned vertically to show the width of the object. The right and left views are aligned horizontally to show the height of the object. The top and right views have the same depth of the object.

Drawing three views:
(1) Analyzing-shape method: Any composite object can be broken into a combination of some primary geometric object. Any of these basic shapes can be positive, classified to the superposition style and the cutting style.
(2) Select the projection: Because the front view is the most important one in the three views, it is very important to select adequate projection direction to form the front view.
(3) Drawing steps: Locate the axis lines, center-lines of symmetry and base lines; Draw the base with H pencil, check the drawing and darken the lines.

Reading the composite views:
Points of reading views: break the object into its individual basic shapes; when reading views at least two views should be read simultaneously, master the meanings of lines and areas in views.

Methods to read views:
(1) Analyzing shape method: Break the object down into its basic geometric solids.
(2) Analyzing lines and planes method: Break the object into various surfaces and lines.
For example, like the following Fig.1.3, the three views of a Composite Object.

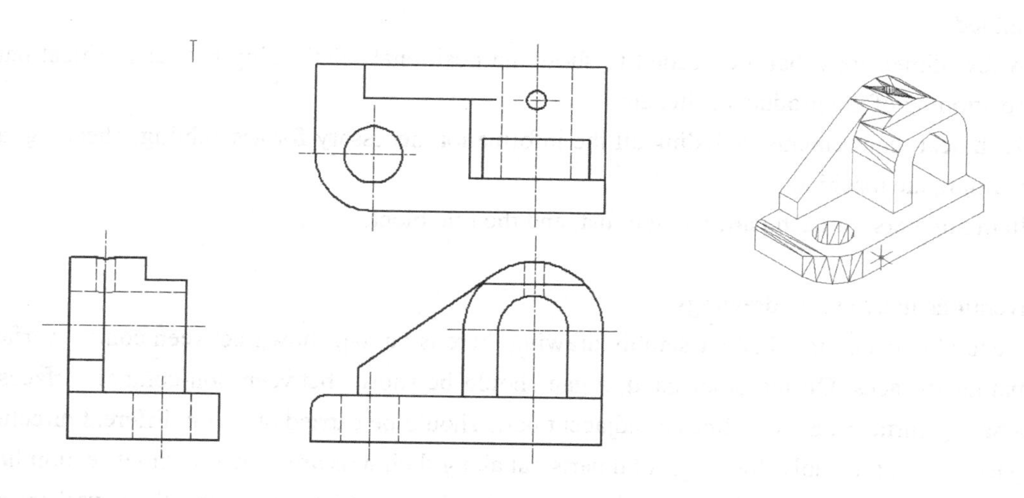

Fig.1.3 Three Views of Composite Object

★ **Detail drawings**

Detail drawings: it is a drawing that indicates the construction, size, and technical requirements of a part. It describes its shape, gives the dimensions, provides all the information needed to make the part.
The contents of a detail drawing:
(1) A sufficient number of views to give a complete shape description of the exterior and interior

constructions of the part.
(2) All the dimensions needed for manufacturing the part.
(3) Technical requirements including tolerances, geometric tolerances, surface roughness, material specification, heat treatments, and so on.

Selection of views: To meet the requirements of making a clear and complete shape description of the object, the first thing to decide before starting to draw is which views are needed and the best way to position the part on the drawing.
(1) Selection of main view: the characteristic shape principle, the functioning position principle, the machining position principle.
(2) Selection of other views: to limit the number of views to the minimum necessary and sufficient principle, to avoid the use of hidden lines principle, to avoid unnecessary repetition of details principle.

★ **Assembly drawings**

Assembly drawings: A drawing that shows the parts of a machine or machine unit assembled in their relative working position is called an assembly drawing.
The contents of an assembly drawing are as follows:
(1) A set of views showing the positional relationship and mutual operation of the parts being assembled.
(2) A few dimensions that are needed to show the positional relationship between critical parts, the positioning of the product at site, etc.
(3) Technical requirements including all the information necessary for assembling, checking, and maintaining the machine.
(4) Item numbers for each part, the item list, and the title block.

Conventions in assembly drawings:
(1) General conventions: In an assembly drawing there is no gap shown between contact surfaces or mating surfaces. On the other hand, a gap should be shown between non-contact surfaces or non-mating surfaces. Section lines of adjacent parts should be carried out with different directions or spaces. In an assembly drawing, solid parts cut along their axis are shown without section lines. Such as shafts, axles, rods, handles, pins, keys, etc. Screws, bolts, nuts, and their washers also keep its shape.
(2) Special conventions: Representation of making the cut along joint face or taking some parts apart, Representation of showing parts separately, Representation of using phantom lines, Exaggerated representation, Simplified representation[3].

Words

graphics	*n.* 制图，图学
drafting	*n.* 草图，制图
drawing	*n.* 绘图，制图，图样
projection	*n.* 投影
dimension	*n.* 尺寸；*v.* 给……标注尺寸
spatial analysis	空间分析
spatial visualization	空间想象
horizontal projection	水平投影
frontal projection	正投影
profile projection	侧投影
quadrant	*n.* 象限
center-lines of symmetry	对称中心线
composite object	组合体
detail drawing	零件图
assembly drawing	装配图
phantom line	假想线

Notes

[1] Graphics comes to our vocabulary from the Greek word grapho, whose extended meaning is *drafting* or *drawing*, the drawing is the primary medium for developing and communicating technical ideas.

【译文】“图学”一词来源于希腊字“grapho”，其延伸意义为“绘图”或“图样”。图样是开发和交流技术思想的主要工具。

[2] According to the Chinese National Standard of Technical Drawings, the first-angle projection is used to make engineering drawings while in some other countries, such as in the USA and Canada, the third-angle projection is used.

【译文】依据中国机械制图国家标准，制图采用第一角投影，而其他一些国家如美国和加拿大则采用第三角投影。

[3] Special conventions: Representation of making the cut along joint face or taking some parts apart, Representation of showing parts separately, Representation of using phantom lines, Exaggerated representation, Simplified representation.

【译文】特殊规定：沿结合面剖切或把某些零件拆开的画法，单独表示零件画法，使用假想线画法，夸大画法，简化画法。

Part B Reading Materials

Similar to an offset in that the cutting-plane line staggers, however, it differs in that the cutting-plane line is offset at some angle other than 90°. When the section is taken the sectional view is drawn as if the cutting—plane is rotated to the plane perpendicular to the line of sight. This is why the right-side sectional view may sometimes be elongated (depending on the shape).

Exercise:

Drawing the front view to aligned section.

The Fig.1.4(a) is the topic and Fig.1.4(b) is the answer.

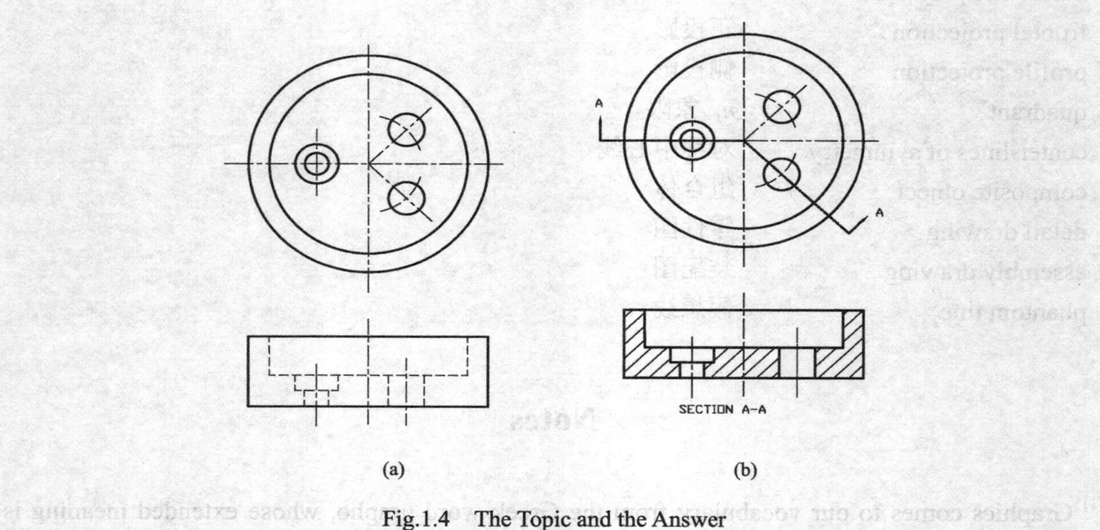

Fig.1.4 The Topic and the Answer

Part C Exercise

A brief introduction to the difference between the first projection and the third projection.

Lesson 2　CAD Application

教学目的和要求

本文简要介绍了计算机辅助设计（CAD）的软件用途及分类，以 AutoCAD 为例，介绍了 AutoDesk 的绘图界面。计算机辅助绘图软件是机械工程专业的通用软件，其操作命令都是以英文输入的。通过本文的学习，可以了解有关 CAD 的英文表达和绘图界面的常见图标的英文名称，有助于学习和应用英文版本 CAD；要求掌握文中所涉及的绘图界面图标的英文名称，并能在实际应用中不断积累绘图命令的英文名称及快捷命令，并能根据文后所附的练习绘制出同样的图形。

重点和难点

（1）重点掌握 CAD 相关的专业术语及表达。
（2）掌握 AutoDesk 绘图界面各部分的名称和图标的名称。

Part A　Text

CAD refers to Computer Aided Drafting. CAD has replaced the traditional drawing instruments such as triangles and pencils, but it still has not, and probably never will, practically replace the ideal freehand sketch or the designer's experience with geometry and graphical conventions and standards.

CAD gives us increased accuracy, productivity and transferability. It affords us the flexibility to change drawings with minimal effort. Before CAD, a minor mistake made on a manual drawing could mean extensive time and cost to correct a major mistake resulted in a recreation of the entire drawing. CAD also eliminates the need to frequently redraw standard equipment, components, and details.

AutoCAD was first released in 1982. AutoCAD is undoubtedly the world's leading CAD software for 2D drafting, detailing, design documentation, and basic 3D design. It is a vector graphics drawing program that uses primitive entities—such as lines, polylines, circles, arcs, and text—as the foundation for blocks and more complex objects. AutoCAD can be customized for the various uses/disciplines by adding interfaces such as Architectural Desktop, AutoCAD Electrical and Mechanical Desktop. AutoCAD has a full set of basic solid modelling and 3D tools, but lacks advanced capabilities of solid modelling applications compared to such software as Pro/ENGINEER and Solid Works[1].

There are many Computer Aided Drafting programs on the market today with each having its own strengths and weaknesses.

CAD distinguishes itself into **three** main categories:

- 2D,
- mid-range 3D solid Modelling,
- high-end 3D hybrid systems.

Some programs are customized for specific disciplines (e.g. ECAD for Electrical Engineering design and MCAD for Mechanical Engineering design). Some programs are designed for entry level drafting, others for advanced design. For example, ArchiCAD has the ability to design virtual tours of building models and also estimate monthly/annual energy cost[2]. Pro/ENGINEER is designed for various platforms, such as UNIX, Windows, and Linux and is exchangeable between platforms with noticeable conversions. AutoDesk Raster Design can aid you in turning scanned images into vector drawing with some editing that you can open with another computer aided drafting program for editing[3].

Interactive third-party software enhances more generic software by allowing information created from one software program to be viewed and edited by another. For example, SolidWorks is a program that can create a 2D drawing from a 3D image and vice versa.

Note:

If this dialog box does not appear when you start AutoCAD your drawing will be set for either imperial or metric, possibly without you knowing which. Therefore, you should set the system variable to a value of 1 so that it will appear when you choose to open a new drawing, as shown in Fig.2.1. To do so, at the command prompt, type **STARTUP** and enter a value of 1.

Another option to ensure you are working in a metric prototype drawing is to select File, New and choose **acadiso.dwt** (metric template).

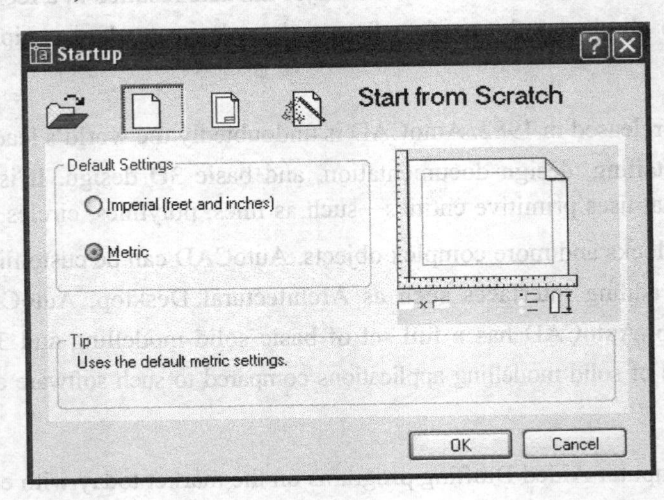

Fig.2.1 Start Up Dialog Box

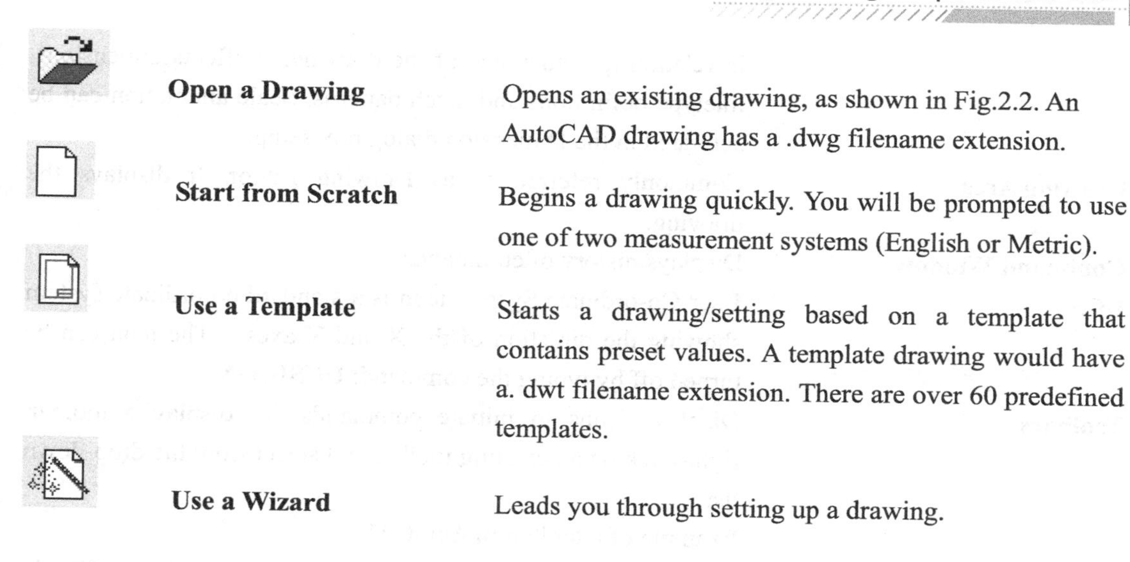

	Open a Drawing	Opens an existing drawing, as shown in Fig.2.2. An AutoCAD drawing has a .dwg filename extension.
	Start from Scratch	Begins a drawing quickly. You will be prompted to use one of two measurement systems (English or Metric).
	Use a Template	Starts a drawing/setting based on a template that contains preset values. A template drawing would have a. dwt filename extension. There are over 60 predefined templates.
	Use a Wizard	Leads you through setting up a drawing.

Note:

Other common filename extensions in addition to those already mentioned include:

- .bak (backup drawing—see saving a drawing).
- .plt (plot file—see plotting).
- .ctb (see plot style table).

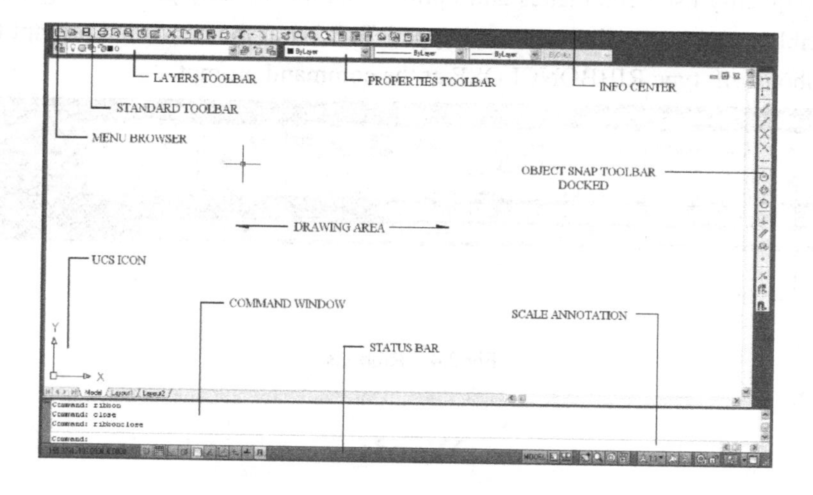

Fig.2.2 Interface (without Ribbons)

Menu Browser Menu commands that are organized in groups such as File, Edit, View, Insert, Format, Tools, Draw, Dimension, Modify, etc.

Status Bar Displays X, Y Co-ordinates and Drawing Aids (SNAP, GRID, ORTHO, POLAR, OSNAP, OTRACK, MODEL). To the right of the Status bar you will see a series of icons used to control the drawings '**Scale Annotations**' to simplify size of annotation in relation to actual size of the drawing. It affects dimensions, linetype definitions and hatch patterns. Scale annotation can be activated in the Dimension dialog box setup.

Drawing Area Commonly referred to as Drawing Editor. It displays the drawing.

Command Window Displays history of commands.

UCS **U**ser **C**o-ordinate **S**ystem icon is a standard co-ordinate system showing the direction of the X and Y axes. The icon can be turned off by typing the command: **UCSICON**.

Toolbars Displays icons to initiate commands. To display a toolbar, right-click on an existing toolbar and select from the drop-down list.

Example of a toolbar in AutoCAD:

Standard Toolbar includes Open, Save, Print, Zoom, Check, Cut, Copy, Paste, etc.

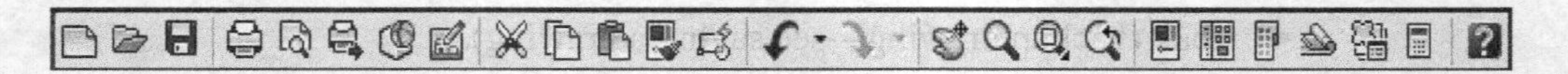

Fig.2.3 Interface (with Ribbons)

Ribbons are frequently used commands and options that are visually grouped together, as shown in Fig.2.4. If ribbons are not visible you may type RIBBON at the command prompt to turn them on. To turn ribbons off, type RIBBONCLOSE at the command prompt.

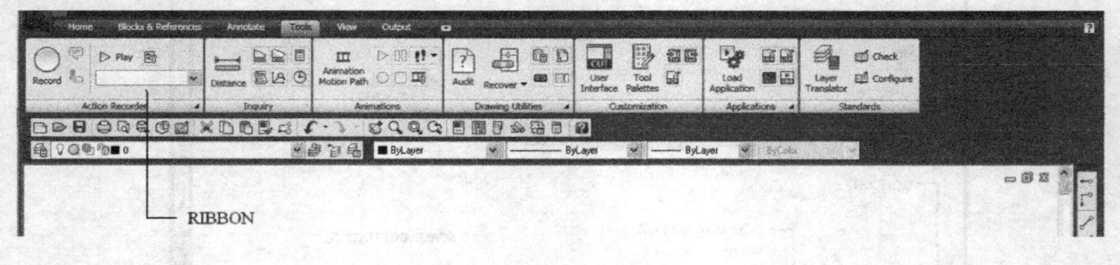

Fig.2.4 Ribbons

Words

vector	*n.* 矢量
graphics	*n.* 制图学；制图法；图表算法；
polylines	*n.* 多段线
circle	*n.* 圆
arc	*n.* 圆弧
block	*n.* 块
imperial	*adj.* 英制的

metric	*adj.* 公制的
dimension	*n.* 尺寸；*vt.* 标出尺寸，尺寸标注
modify	*n.* 修改
linetype	*n.* 线型
activate	*vt.* 激活
dialog box	*n.* 对话框
co-ordinate	*n.* 坐标
icon	*n.* 图标
paste	*vt.* 粘贴

Notes

[1] AutoCAD has a full set of basic solid modelling and 3D tools, but lacks advanced capabilities of solid modelling applications compared to such software as Pro/ENGINEER and SolidWorks。

【注释】

①Pro/ENGINEER（简称 Pro/E）：该操作软件是美国参数技术公司(PTC)旗下的 CAD/CAM/CAE 一体化的三维软件。Pro/E 第一个提出了参数化设计的概念，并且采用了单一数据库来解决特征的相关性问题。另外，它采用模块化方式，用户可以根据自身的需要进行选择，而不必安装所有模块。应用 Pro/E 提供的基于特征方式，能够把从设计到生产全过程集成，实现并行工程设计。它不但可以应用于工作站，而且也可以应用到单机上。Pro/E 采用了模块方式，可以分别进行草图绘制、零件制作、装配设计、钣金设计、加工处理等，保证用户可以按照自己的需要进行选择使用。

②SolidWorks：SolidWorks 是达索系统(Dassault Systemes S.A)下的子公司，专门负责研发与销售机械设计软件的视窗产品，公司总部位于美国马萨诸塞州。SolidWorks 软件功能强大，组件繁多，有功能强大、易学易用和技术创新三大特点，这使得 SolidWorks 成为领先的、主流的三维 CAD 解决方案。SolidWorks 能够提供不同的设计方案、减少设计过程中的错误以及提高产品质量。SolidWorks 不仅提供如此强大的功能，而且对每个工程师和设计者来说，操作简单方便、易学易用。在目前市场上所见到的三维 CAD 解决方案中，SolidWorks 是设计过程比较简便而方便的软件之一。美国著名咨询公司 Daratech 评论：“在基于 Windows 平台的三维 CAD 软件中，SolidWorks 是最著名的品牌，是市场快速增长的领导者。”

[2] Some programs are customized for specific disciplines (e.g. ECAD for Electrical Engineering design and MCAD for Mechanical Engineering design). Some programs are designed for entry level drafting, others for advanced design. For example, ArchiCAD has the ability to design virtual tours of building models and also estimate monthly/annual energy cost.

【注释】

① ECAD：电气工程计算机辅助设计。

② MCAD：机械工程计算机辅助设计。

③ ArchiCAD：虚拟建筑设计，ArchiCAD 不同于原始的二维平台及其他三维建模软件，其

中最重要的一点就是能够利用 ArchiCAD 平台创建的虚拟建筑信息模型，进行高级解析与分析，如绿色建筑的能量分析、热量分析、管道冲突检验、安全分析等。在 2D 环境下，每一张图样都是建筑师一笔一笔地以线条形式画出来的：先从平面图开始绘制，然后画立面图、剖面图，最后再按照项目进展更改所有的图样。然而，在“虚拟建筑”中，ArchiCAD 虚拟建筑设计平台将彻底地改变这个工作过程，使建筑师能够从各种繁杂的图样绘制工作中抽身。在这里，设计的核心工作不再是绘制施工图样，而是以虚拟建筑信息模型为工作中心，所有的图样都直接从模型中生成。建筑师将集中更多的精力用于建筑物的设计，而图样将彻底成为设计的副产品。

[3] AutoDesk Raster Design can aid you in turning scanned images into vector drawing with some editing that you can open with another computer aided drafting program for editing.

【注释】

①AutoDesk：中文名称为欧特克，是全球最大的二维、三维设计和工程软件公司，旗下拥有 AutoCAD、3DS Max 等知名软件。

②Raster Design：译为光栅设计软件。在 AutoCAD 环境中，通过使用 Autodesk Raster Design 软件的光栅编辑、操纵和矢量化工具，扩展光栅数据的价值。在项目中添加扫描的图纸、地图、航拍照片、卫星图像和其他形式的光栅数据，并使用功能强大的清理工具来提高图像质量。可以仅对那些需要修改的光栅图元进行矢量化，从而节省编辑时间。

Part B Reading Materials

Viewports

When working in Model space you draw geometry in tile viewports [Fig.2.5(a)] which are represented by Viewport Table Record objects. You can display one or several different viewports at a time. If several tiled viewports are displayed, editing in one viewport affects all other viewports. However, you can set the magnification, viewpoint, grid, and snap settings individually for each viewport.

In paper space, you work in floating viewports [Fig.2.5(b)] which are represented by viewport objects and can contain different views of your model. Floating viewports are treated as objects that you can move, resize, and shape to create a suitable layout. You also can draw objects, such as title blocks or annotations, directly in the Paper space view without affecting the model itself.

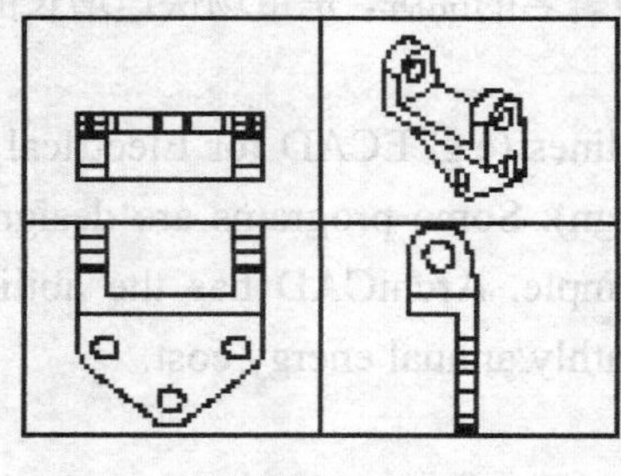

(a) Tiled Viewports

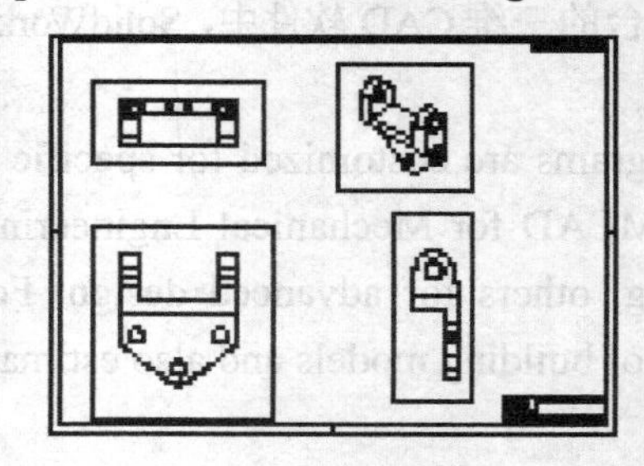

(b) Floating Viewports

Fig.2.5 Tiled Viewports and Floating Viewports

Part C Exercises

Can you draw the figure of hollow pipe below according to the drawing process below? Please explain the meanings of the following commands.

1. surfab 2:
2. tabsurf:
3. move:
4. revsurf:
5. hide:

Command: surftab 2
New current value for surftab 2 (2 to 32766) <6>: 10
Command: _tabsurf
Select entity to extrude:
Select line or open polyline for extrusion path:
Command: move
Select objects:
Specify base point or [Displacement] <Displacement>:0,0,1000
Command: _revsurf
Select a linear entity to revolve:
Select the axis of revolution:
Angle to begin surface of revolution <0>:
Degrees to revolve entity (+ for ccw, - for cw <360>: 90
Command: _revsurf
Select a linear entity to revolve:
Select the axis of revolution:
Angle to begin surface of revolution <0>:
Degrees to revolve entity (+ for ccw, - for cw <360>: -90
Command: _move
Select objects:
Specify base point or [Displacement] <Displacement>:0,0,1000
Command: hide

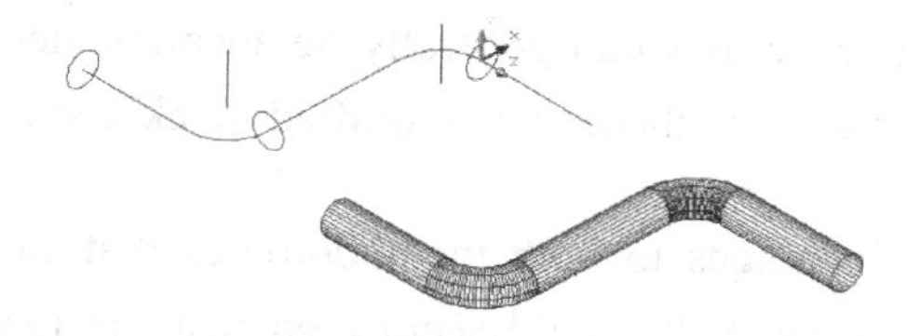

Fig.2.6 Hollow Pipe

Lesson 3 UG Application

教学目的和要求

本文介绍 UG 软件提供的混合实体/曲面建模的功能优势，首先说明实体建模和曲面建模的优缺点，以及为了实现这两种建模的优势组合而产生的混合建模；其次说明混合建模的应用场合并举例说明其特点。在熟练应用 UG 软件的基础上，再通过本文的学习，读者可以了解 UG 软件中关于混合实体/曲面建模的应用及相关术语的英文表达。

重点和难点

（1）重点掌握 UG 软件中关于建模的相关专业术语及表达。

（2）本文的学习首先建立在对 UG 软件熟悉的基础上，有了基础才能更好地理解本文，建议有选择地学习。

Part A Text

What is hybrid solid/surface modeling and what are the benefits?

Almost all Modelers today use a boundary representation to represent solids. In a nutshell, this means that the object is represented by the surfaces that define the outer boundary of the volume. Thus, the axiomatic principle that is key to an understanding of solid versus surface modeling is that given enough time and the proper set of tools, any object that can be modeled with solids can be also modeled with surfaces[1]. The converse however is not true. In order to provide a reasonable set of operations, solid Modelers typically restrict the user to a finite set of shapes (i.e. spheres, boxes, etc.) and/or procedures (i.e. extrude, revolve, sweep, fillet etc.) that can be used to define the geometry of the object. This inherently limits the range of objects that can be described. In addition, they do not allow the representation of zero thickness partitions, such as the parting surface of a Mold. Because solid Modelers require that a true bounded volume be described at all times, manipulation and *fine-tuning* of the individual bounding faces is generally restricted.

Why then, with these limitations, has the use of solid modeling grown so dramatically? The primary reason is productivity. Since solid Modelers can perform in a single step what would generally take numerous operations with surfaces, most objects can be described much more efficiently. Also, these solid operations can generally be accompanied by describing dimensions and constraints in a fashion that allow them to be modified quickly and easily.

So, we see that there are advantages to both methodologies that can be summarized as speed (solids) versus flexibility (surfaces). It would seem then that the ideal Modeler would provide both techniques, in an environment that allows the user to easily switch back and forth between them. This is known as hybrid solid/surface modeling. In practice, the synergy of this combination has proven to exceed the sum of its parts. A well-designed hybrid modeling system allows the user to leverage the power and efficiency of solids whenever possible, yet never force a design to be compromised for lack of an appropriate tool to achieve the desired geometry.

What applications is hybrid solid/surface modeling best suited for?

The ability to have a single CAD/CAM system that can be fast enough to design a gearbox on Monday and flexible enough to design a body panel on Tuesday is perhaps a compelling enough reason to have integrated solid and surface design. However, to truly realize the benefits of integrated solid/surface modeling one must look beyond this either/or paradigm. Consider, for example, the design of a camera or cellular phone. A true hybrid modeling system provides the functionality to design a beautiful, ergonomic body for the appliance with surface modeling tools and in the same environment apply a powerful and intelligent set of solid modeling tools to convert it into a thin shelled case, split it in half and add all the required bosses, ribs and flanges. If requirements dictate a geometric feature that cannot be achieved with one of the standard solid feature tools, in a hybrid modeling system surfaces can be used to *knife and fork* the desired shape[2]. Beyond the design of products that don't easily fit the standard feature-modeling paradigm, there are several other applications where the integration of solids and surfaces yields great benefits. One of these is for the repair or modification of models that have geometry problems, most often as a result of data translation or Modelers that allow geometry inaccuracies to accumulate.

Another is for the design of Molds & Dies, tools or fixtures where a shape needs to be partitioned into pieces along irregular boundaries, or faces of an object need to be extended or manipulated. Quite often supplemental geometry needs to be created for the purposes of NC tool path creation. A CAM system with an integrated hybrid Modeler is a tremendously powerful tool.

Will your organization benefit from Hybrid modeling?

Does your company do NC programming?

If yes, you need hybrid modeling. If no, continue.

Does your company work with imported solid or surface data via IGES, STEP or VDA[3]? If yes, you need hybrid modeling.

If no, continue.

Does your company design parts that have free-form shapes? Any feature of the part should be considered and not just the overall shape. For example, your design is 98% prismatic in shape but has one feature such as a cam that requires a free form or variably swept face.

If yes, you need hybrid modeling.

If no, continue.

Does your company design Molds, Dies, tools, or fixtures? If yes, you need hybrid modeling.

If no, continue.

Does your company use a solid Modeler that is limited in the functions available to create your designs? For example, does your parts have complex fillets or corner blends that your *state-of-the-art* solid Modeler is not capable of creating?

If yes, you need hybrid modeling.

If no, continue.

Are your designs primarily surface modeling based but you would like to realize the benefits of parametric, dimension driven modeling?

If yes, you need hybrid modeling.

Hybrid modeling example 1. Open the file *Advanced_Modeling* which is located on the training CD. Edit the part *01_Hybrid1*. Step through the part history to see how the model was constructed. Notice how the Bottom Face has been deliberately removed in order to make this a surface model or open shape.

In spite of this you can see that the extruded cut operations work even though the part is not a solid. In wire frame mode, some of the edges of the faces are shown as blue and dotted to indicate that they are not attached to other face edges. As shown Fig.3.1(a).

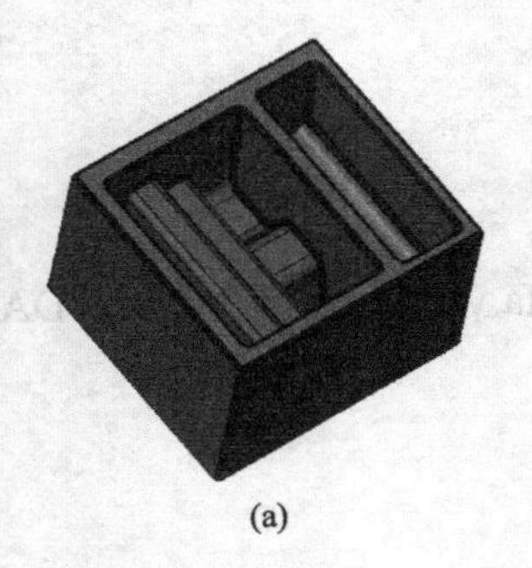
(a)

(b)

Fig.3.1 Wire Frame Mode

In shaded mode ZW3D will show you the inside (or negative normal) faces clearly in a pink color. When creating faces that share common edges, the normals will be oriented in the same direction automatically. If the faces eventually form an enclosed volume (i.e. a solid) then all the faces will be orientated so that the positive normal points away from the solid body.

Rotate the view so that you are looking underneath the part as shown below Fig.3.1(b). Note: open the face attribute option under Attributes pull-down menu, then set the color of the front and back from the pop-up dialog box.

You can clearly see that the extruded faces and the main body need to be trimmed to each other. Although there are many powerful trimming options in ZW3D that allow the user to trim the faces, it is nevertheless a time-consuming process. Instead we will use hybrid modeling techniques to quickly join the Shapes to each other.

From the Shape Tab, using the Combine > Add command, join Pick the main body as the base shape. Middle-click to advance the menu. With ZW3D hybrid modeling the user can use commands, normally associated with Solid Modelers, on surfaces and or solids.

For example. Using the fillet command let's put a 5mm fillet on all of the edges (window pick the entire part). Remember all of the commands shown in this part were done on a surface model. No need to save the file.

Words

hybrid modeling	*n.* 混合模型
modeler	*n.* 模型师；建模程序
nutshell	*n.* 简言之，一言以蔽之
axiomatic	*adj.* 不证自明的；原则的
extrude	*v.* 拉伸
revolve	*v.* 旋转
sweep	*v.* 扫掠
fillet	*v.* 圆角
inherently	*adv.* 固有地
manipulation	*n.* 操作
fine-tuning	微调
paradigm	*n.* 范式
ergonomic	*adj.* 人类环境改造学的；符合人机工程学的
boss	*n.* 凸台
rib	*n.* 筋
flange	*n.* (机械等的)凸缘，(火车的)轮缘

Notes

[1] In a nutshell, this means that the object is represented by the surfaces that define the outer boundary of the volume. Thus, the axiomatic principle that is key to an understanding of solid versus surface modeling is that given enough time and the proper set of tools, any object that can be modeled with solids can be also modeled with surfaces.

【注释】In a nutshell：简言之，一言以蔽之。

【译文】简言之，即由确定了该体积外边界的多个曲面来表达某一对象。因而，对实体曲面建模的认识较为关键的原则，就是在时间充分、工具组合适的前提下，对那些能利用实体进行建模的对象，都可以用若干曲面进行建模。

[2] If requirements dictate a geometric feature that cannot be achieved with one of the standard solid feature tools, in a hybrid modeling system surfaces can be used to *knife and fork* the desired shape.

【注释】knife and fork：分割并构建。

【译文】若所要求的几何体特征无法通过其中一种标准实体特征工具达成，则可通过混合建模系统，利用若干曲面来“分割并构建”所需的造型。

[3] Does your company work with imported solid or surface data via IGES, STEP or VDA?

【注释】

① IGES：全称是 The Initial Graphics Exchange Specification，即初始化图形交换规范，它基于 Computer-Aided Design (CAD)&Computer-Aided Manufacturing (CAM) systems (计算机辅助设计和计算机辅助制造系统)不同计算机系统之间通用的 ANSI 信息交换标准。

② STEP：全称是 Standard for the Exchange of Product Model Data，即产品模型数据交互规范。这个标准是国际标准化组织制定的用于描述整个产品生命周期内产品信息的标准，是一个正在完善中的“产品模型数据交换标准”。它是由国际标准化组织(ISO)工业自动化与集成技术委员会(TC184)下属的第四分委会(SC4)制定的，ISO 正式代号为 ISO-10303。它提供了一种不依赖具体系统的中性机制，旨在实现产品数据的交换和共享。它不仅适合于交换文件，也适合于作为执行和分享产品数据库和存档的基础。发达国家已经把 STEP 标准推向工业应用，它的应用显著降低了产品生命周期内的信息交换成本，提高了产品研发效率，成为制造业进行国际合作、参与国际竞争的重要基础标准，是保持企业竞争力的重要工具。

③ VDA：全称是 Video Display Adapter，即视频显示适配器。VDA 文件是一个 VDA-FS CAD 文档，而 VDA-FS 是一种用于将表面模型从一个 CAD 系统传输到另一个 CAD 系统的 CAD 数据交换格式。

【译文】贵公司是否使用通过 IGES、STEP 或VDA 导入的实体或曲面数据？

Part B Reading Materials

Shell (Shape Tab)

Use this command to create an offset shell feature (Fig.3.2) from a solid. The offset will apply to all faces of the solid except those that have Face offset attributes. You can also specify which faces of the shell if any should remain open.

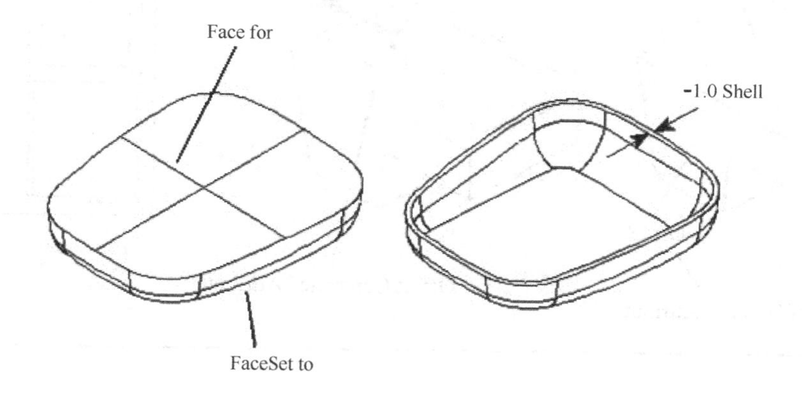

Fig.3.2 Shell

1) Shell Solid

Shells can also be applied to open shape features.

A negative value for thickness creates the wall thickness inside the original shape (on the pink side).

Firstly, Optional inputs.

Secondly, Create side faces.

Use this command option to control the creation of side faces during the Offset Face and Shell commands. When you select side faces the shape will be reconnected to form a closed solid (recommended).

Finally, Intersections.

Use this command option to check for self-intersecting faces during the Shell commands. This option will also detect disappearing fillet faces, fillet corners, and chamfers. These checks require some time to complete depending on part complexity.

Part C　Exercises

Please translate the English annotation in the Fig.3.3 below.

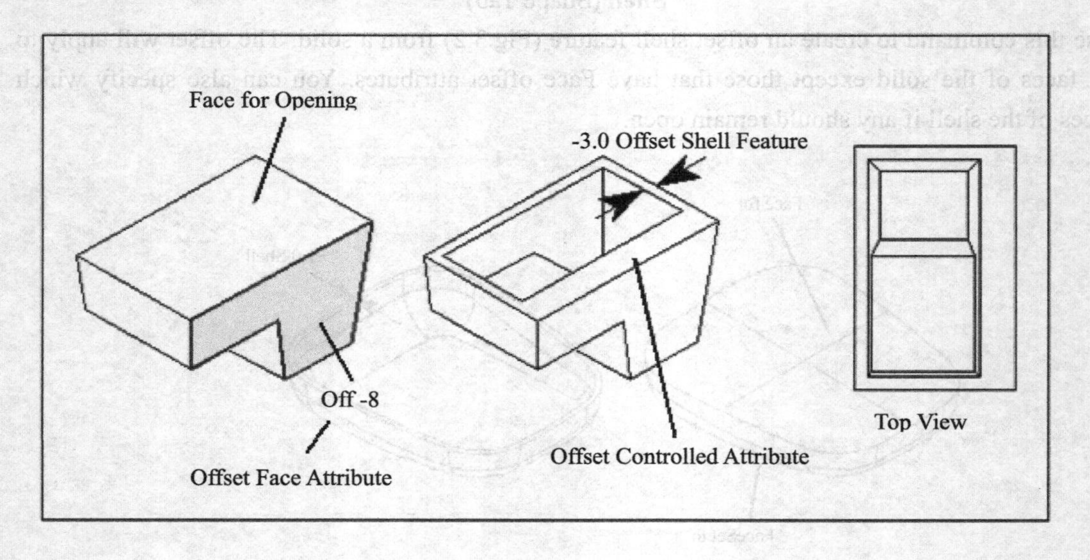

Fig.3.3　Drawing Process

Lesson 4 Pro/ENGINEER Application

教学目的和要求

本文为Pro/ENGINEER（简称Pro/E）的实例应用，简单介绍Pro/ENGINEER拉伸的一般步骤和操作，描述Pro/ENGINEER设计过程，并以夹钳体为实例，简单介绍Pro/ENGINEER实体建模中的界面、操作和命令。通过本文的学习，可以了解有关Pro/ENGINEER的英文表达和常见功能操作、命令术语的英文名称，有助于英文版本的Pro/ENGINEER学习和应用；要求掌握文中所涉及的Pro/ENGINEER中各种术语的英文名称，且在实际应用中不断积累各类相关词汇的英文名称，并能对文后所附的练习进行理解和翻译。

重点和难点

（1）重点掌握Pro/ENGINEER相关的专业术语及表达。

（2）掌握Pro/ENGINEER实体建模的一般操作步骤、命令和方法。

Part A Text

Extrusions

The design of a part using Pro/E starts with the creation of base features (normally datum planes), and a solid protrusion. Other protrusions and cuts are then added in sequence as required by the design[1]. You can use various types of Pro/E features as building blocks in the progressive creation of solid parts (Fig.4.1). Certain features, by necessity, precede other more dependent features in the design process. Those dependent features rely on the previously defined features for dimensional and geometric references.

★ Solid Modeling

The progressive design of features creates these dependent feature relationships known as parent-child relationships. The actual sequential history of the design is displayed in the Model Tree[2]. The parent-child relationship is one of the most powerful aspects of Pro/E and parametric modeling in general. It is also very important after you modify a part. After a parent feature in a part is modified, all children are automatically modified to reflect the changes in the parent feature. It is therefore essential to reference feature dimensions so that Pro/E can correctly propagate design modifications throughout the model.

Fig.4.1 Clamp

An extrusion is a part feature that adds or removes material[3]. A protrusion is always the first solid feature created. This is usually the first feature created after a base feature of datum planes. The Extrude Tool is used to create both protrusions and cuts. A tool chest button is available for this command or it can be initiated using Insert Extrude from the menu bar.

★ **The Design Process**

It is tempting to directly start creating models. Nevertheless, in order to build value into a design, you need to create a product that can keep up with the constant design changes associated with the design-through-manufacturing process. Flexibility must be *built in* to the design. Flexibility is the key to a friendly and robust product design while maintaining design intent, and you can accomplish it through planning. To plan a design, you need understand the overall function, form, and fit of the product. This understanding includes the following points:

- Overall size of the part.
- Basic part characteristics.
- The way in which the part can be assembled.
- Approximate number of assembly components.
- The manufacturing processes required to produce the part.

★ **Clamp**

The clamp in Fig.4.2 is composed of a protrusion and two cuts. A number of things need to be established before you actually start modeling. These include setting up the environment, selecting the units, and establishing the material for the part.

Before you begin any part using Pro/E, you must plan the design. The design intent will depend on a number of things that are out of your control and on a number that you can establish. Asking yourself a few questions will clear up the design intent you will follow: Is the part a component of an assembly? If so, what surfaces or features are used to connect one part to another? Will geometric tolerance be used on the part and assembly? What units are being used in the design, SI or decimal inch? What is the part's material? What is the primary part feature? How should I model the part, and what features are best used for the primary protrusion (the first solid mass)?

On what datum plane should I sketch to model the first protrusion? These and many other questions will be answered as you follow the systematic lesson part. However, you must answer many of the questions on your own when completing the lesson project, which does not come with systematic instructions.

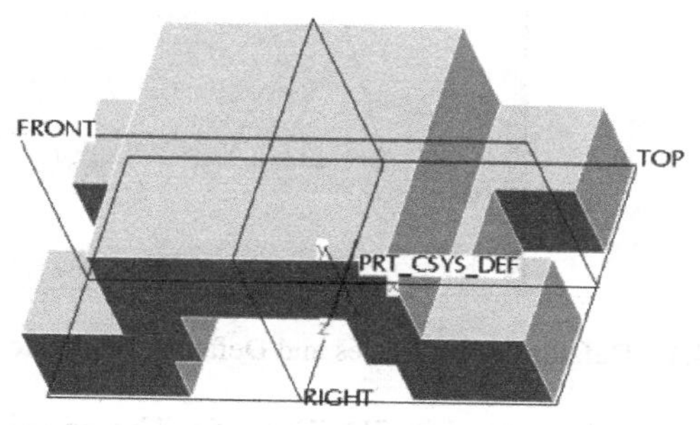

Fig.4.2 Clamp and Datum Planes

The material file, STEEL, is without any file information (Fig.4.3). As an option, if your instructor provides you with the specifications, or you are familiar with setting up material specs, you can edit the file using: Edit Setup Material Edit Steel Accept fill in the information File Save File Exit Done.

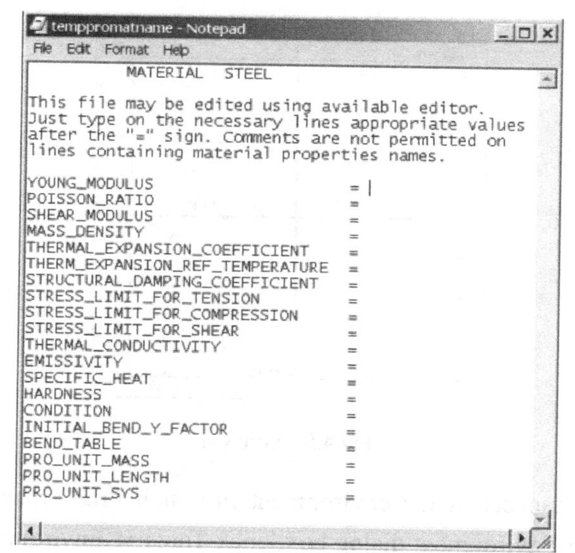

Fig.4.3 Material File

Since ☑ Use default template was selected, the default datum planes and the default coordinate system are displayed in the graphics window and in the Model Tree (Fig.4.4). The default datum planes and the default coordinate system will be the first features on all parts and assemblies. The datum planes are used to sketch on and to orient the part's features. Having datum planes as the first features of a part, instead of the first protrusion, gives the designer more flexibility during the design process. Picking on items in the Model Tree will highlight that item on the model.

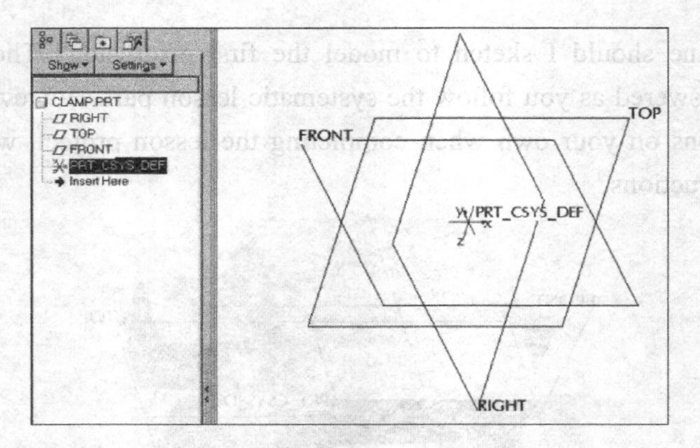

Fig.4.4 Default Datum Planes and Default Coordinate System

The sketch is now displayed and oriented in 2D (Fig.4.5). The coordinate system is at the middle of the sketch, where datum RIGHT and datum TOP intersect. The X coordinate arrow points to the right and the Y coordinate arrow points up. The Z arrow is pointing toward you (out from the screen). The square box you see is the limited display of datum FRONT. This is similar to sketching on a piece of graph paper. Pro/E is not coordinate-based software, so you need not enter geometry with X, Y, and Z coordinates as with many other CAD systems.

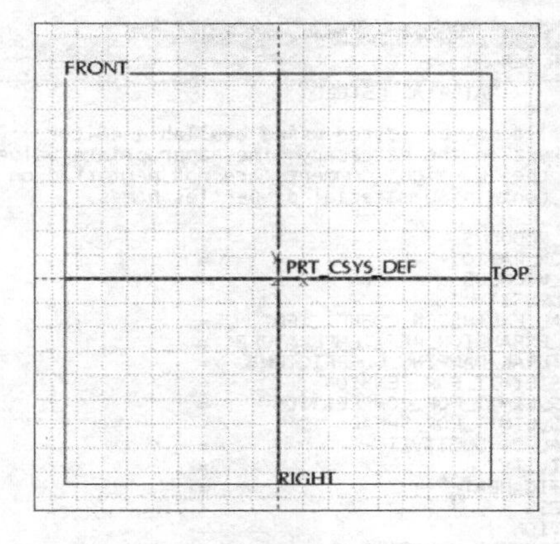

Fig.4.5 Grid On

You can control many aspects of the environment in which Pro/E runs with the Environment dialog box. To open the Environment dialog box, click Tools ⇒ Environment on the menu bar or click the appropriate icon in the toolbar. When you make a change in the Environment dialog box, it takes effect for the current session only. When you start Pro/E, the environment settings are defined by your configuration file, if any otherwise, by Pro/E configuration defaults[4].

Words

datum	*n.* 基点，基线，基面
protrusion	*n.* 伸出，突出；加材料

progressive	*adj.* 进步的；不断前进的；进行的
sequential	*adj.* 按次序的，相继的
propagate	*vt.* 传播；使遗传；扩散，使蔓延
toolchest	*n.* 工具箱
propagate	*vt.* 传播；传送；繁殖；宣传
tolerance	*n.* 公差
systematic	*adj.* 系统的，有规则的
spec	*n.* 规格；说明书
orient	*vt.* 标定方向；以……为参照
stationary	*adj.* 不动的；固定的；静止的
configuration	*n.* 配置；布局，构造

Notes

[1] The design of a part using Pro/E starts with the creation of base features (normally datum planes), and a solid protrusion. Other protrusions and cuts are then added in sequence as required by the design.

【注释】

① Pro/ENGINEER（简称 Pro/E）作为当今世界机械 CAD/CAE/CAM 领域的新标准而得到业界的认可和推广，是现今主流的 CAD/CAM/CAE 软件之一，特别是在国内产品设计领域占据重要位置。Pro/ENGINEER 2001，Pro/ENGINEER 2.0，Pro/ENGINEER 3.0，Pro/ENGINEER 4.0，Pro/ENGINEER 5.0，Creo 1.0，Creo 2.0，Creo 3.0 等都是指 Pro/ENGINEER 软件的版本。

② datum planes：基准面。基准面主要用于绘图平面、参照面、参照基准和确定方向等。

protrusion：加材料，是 Pro/ENGINEER 中实体建模的一种操作方法。

cut：减材料，是 Pro/ENGINEER 中去除材料的一种操作方法。

[2] The progressive design of features creates these dependent feature relationships known as parent-child relationships. The actual sequential history of the design is displayed in the Model Tree.

【注释】parent-child relationships：父子关系。父子关系实际是指两个特征的依赖关系，例如，如果后一个特征是在前一个特征的一个表面上进行拉伸的，那么后一个特征就是依赖于前一个特征而存在的。这时候，这两个特征就具有父子关系，前一个特征是父特征，后一个特征是子特征。如果前一个特征创建失败，那么将会导致后一个特征也会失败。建立正确的父子关系的一个基本原则就是创建特征时只使用必需的参考。

[3] An extrusion is a part feature that adds or removes material.

【注释】Extrusion：拉伸。拉伸是 Pro/ENGINEER 中定义三维几何和零件建模的一种方法，通过将二维截面延伸到垂直于草绘平面的指定距离处来实现。

[4] When you start Pro/E, the environment settings are defined by your configuration file, if any otherwise, by Pro/E configuration defaults.

【注释】

① 该句为省略句，在“by Pro/E configuration defaults”之前省略了“the environment settings are defined”。

② environment settings：环境变量设置。Pro/ENGINEER 中的环境变量决定了零部件设计时的操作环境，如根据什么标准进行设计、采用什么单位制、计算精度、文字高度、公差标准等，一般对默认的环境变量需要进行重新配置。

Part B Reading Materials

You can create drawing templates that help you create drawings automatically with the new drawing templates. Use them to define the layout of views, set view display, place notes, define tables, create snap lines, and show dimensions. You can create customized drawing templates for different types of drawings. For example, you could create a template for a machined part versus a cast part. The machine part template could define the views that are typically placed, set the view display of each view (that is, show hidden lines), place company standard machining notes, and automatically create snap lines for placing dimensions. Drawing templates are used when creating a drawing and automatically create the views, set the desired view display, create snap lines, and show model dimensions based on the template.

The drawing templates improve efficiency and productivity by allowing you to create portions of drawings automatically.

★ Procedure

Click File > New. The New dialog box opens.

Click Drawing, and then type the name of the template you are creating or accept the default.

Clear the Use default template checkbox, and then click OK. The New Drawing dialog box opens.

Click Empty or Empty with format, and then specify the orientation of the template by clicking Portrait, Landscape, or Variable.

Specify the size of the template, and then click OK.

In the Applications menu, click Template to enter Drawing template mode, and then click Views > Add Template. The Template View Instructions dialog box opens.

Type the View Name or accept the default, and then specify the View Orientation.

In the Model *Saved View Name text box, orient the view.*

Specify view options and view values in the View Options and View Values areas.

Click Place View and select the location of the General view.

Note: After you place the view, you now have the options to move the symbol, edit the view symbol, or to replace the view symbol.

To place additional views, click New, type the new view name, and orient the new view. Specify the view options and view values of the new view.

When you are done placing all of the desired views, click OK. Save the template.
Break line or use one of these standard break lines to save time.

Part C Exercises

Translate the following paragraphs into Chinese:

Because some settings will not activate until Pro/E is restarted, many users will exit Pro/E after making changes to their conFig.pro file and then restart, just to make sure the settings are doing what they are supposed to. Do that now. This is not quite so critical since the Options window shows you with the lightning/wand/screen icons whether an option is active. However, be aware of where Pro/E will look for the conFig.pro file on start-up, as discussed above. If you have saved conFig.pro in another working directory than the one you normally start in, then move it before starting Pro/E. On the other hand, if you have settings that you only want active when you are in a certain directory, keep a copy of conFig.pro there and load it once Pro/E has started up and you have changed to the desired directory.

To keep things simple, and until you have plenty of experience with changing the configuration settings, it is usually better to have only one copy of conFig.pro in your startup directory. Note that it is probably easier to make some changes to the environment for a single session using Tools > Environment. Also, as is often the case when learning to use new computer tools, don't try anything too adventurous with conFig. pro in the middle of a part or assembly creation session — you never know when an unanticipated effect might clobber your work!

Lesson 5 SolidWorks Application

教学目的和要求

本文介绍 SolidWorks 软件的一些简单用法。运用实例和图片，描述了 SolidWorks 建模中模板的启用、原点的选择、对称使用、尺寸命名、方程引用、参照创建、圆角倒角以及特征库等操作命令。通过本文的学习，读者可以了解有关 SolidWorks 的英文表达和常见操作命令术语的英文名称，有助于学习和应用英文版本 SolidWorks；要求掌握文中所涉及的 SolidWorks 关于建模的英文术语名称，并能在实际应用中不断地积累建模操作命令的英文名称，能对文后所附的练习进行理解和翻译。

重点和难点

（1）重点掌握 SolidWorks 相关的专业术语及表达。
（2）掌握 SolidWorks 建模的一般操作步骤和方法。

Part A Text

Best practices are simply ways of bringing about better results in easier, more reliable ways. By incorporating some best practices into your SolidWorks modeling routine, you can streamline your design work and focus on your design instead of driving SolidWorks[1]. Below you will find some of the best practices that used when creating part models in SolidWorks.

★ Start with a good template

Creating templates (Fig.5.1) is one of the first things you should do before starting production work (or when starting a new project, if required). The part template provides the foundation that all your models will be built upon. This is especially important if working with other SolidWorks users on the same project, it will ensure consistency across the project.

The template can determine which units and drafting standards are used, and also contains any project-specific custom properties that are desired. It is helpful to keep all metadata associated with part and assembly files (such as standard tolerances and most of the other data that gets read into drawing title blocks) in the respective files themselves. This way, all information related to that particular file is embedded in the custom properties, and the file can stand on its own if, for example, you are dimensioning the part in the file itself.

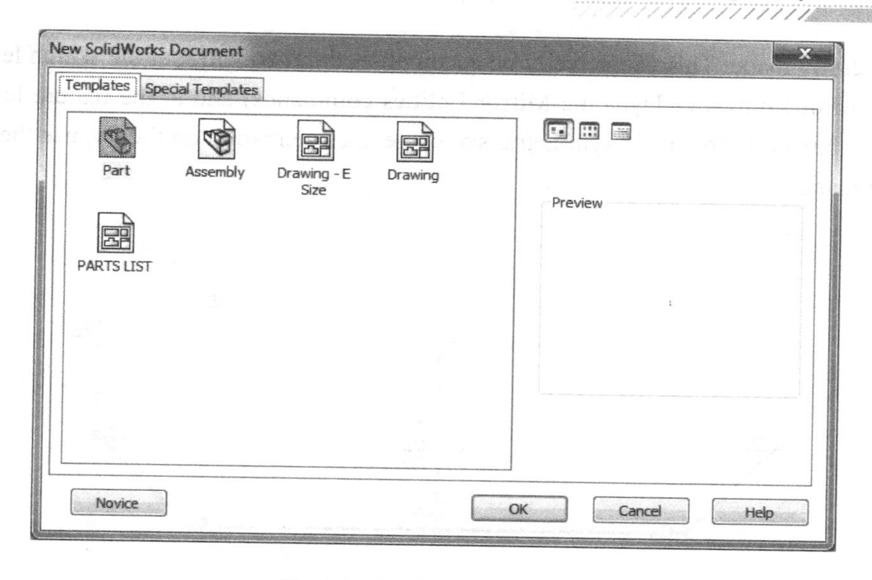

Fig.5.1 SolidWorks Templates

★ **Model parts about the origin**

By always starting your part models at the part's origin (Fig.5.2), you're starting at a known reference that is readily accessible in other environments (i.e., other parts, assemblies, and drawings). This also creates consistency, which is a good thing — especially if you are working with other SolidWorks users.

Sure, you can create your models anywhere you like and fix a point in space, but if the initial sketch is not related to the part's origin, you should be aware of potential issues that may reveal themselves as the design evolves. For instance, if you have features that you want to be symmetrical, you may have to create additional reference geometry in order to mirror.

The only time you overlook this is when doing top-down modeling (working in-context in an assembly). When creating a new part in-context, the part origin is located at the assembly origin. In my experience, that is generally a good thing when creating parts in this manner.

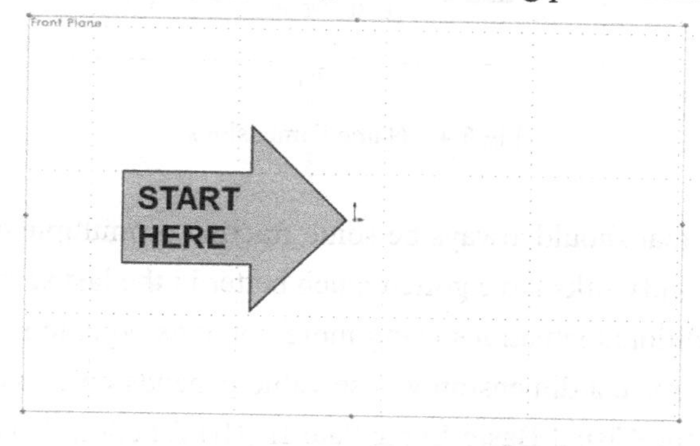

Fig.5.2 Starting with Origin

★ **Use symmetry whenever possible**

Using symmetry (Fig.5.3) wherever possible is a way to simplify your sketches, allowing changes to be more easily made. This can actually be accomplished in two ways: at the sketch level (using either the Mirror Entities or Dynamic Mirror Entities commands) and at the feature level (using the Mirror command). To enable symmetric sketch geometry creation on the fly, use the Dynamic Mirror command.

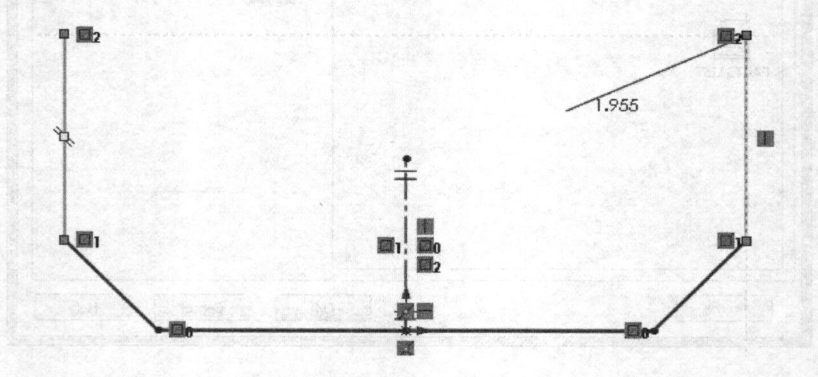

Fig.5.3 Use Symmetry

★ **Name dimensions for reference**

If you are going to be using Equations or building Design Tables, make things a little easier on yourself (and your co-workers) by naming your dimensions (Fig.5.4) as you create them. This is an especially handy tip when modeling configurations of a commercial part from a specification where the dimensions in a table correspond to alphabetical dimensions on a schematic.

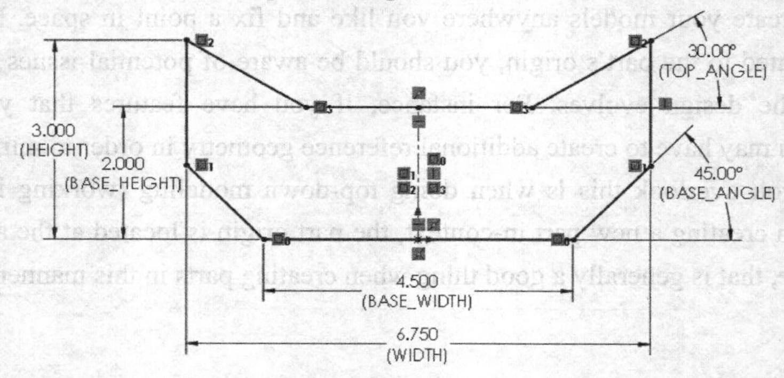

Fig.5.4 Name Dimensions

★ **Use equations**

If you have a dimension that should always be some fraction or multiple of another, then model it that way! Equations in SolidWorks have gotten much better in the last several releases.

You can also create conditional situations using more complex equations (Fig.5.5) and VBA. For instance, if you want to create a dimension whose value depends on a range of values for another dimension, you can use the Visual Basic Immediate If (IIf) function[2]. The IIf function takes the form of IIf(eval, then, else). To put it in SolidWorks format:

D2@ Sketch1 = (IIf (*D1@Sketch1*>6 AND *D1@Sketch1*<12, 8, 4))

In this circumstance, *D2@Sketch1* would evaluate to 8 for values of *D1@Sketch1* between 6 & 12, and 4 for all other values.

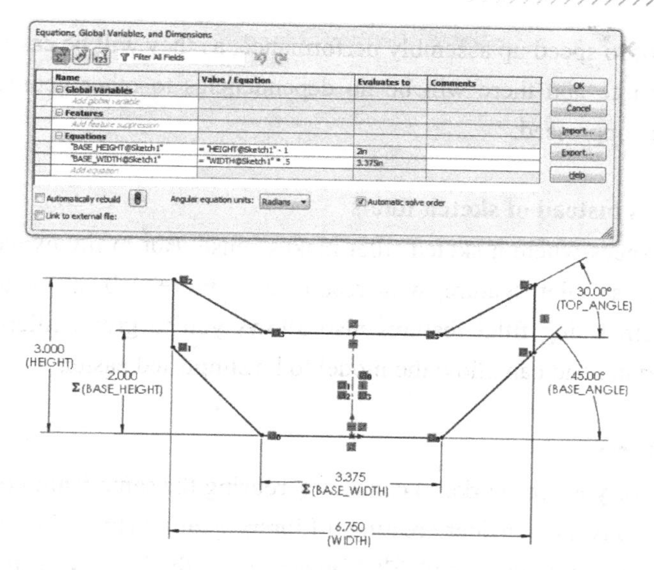

Fig.5.5 Use Equations

★ **Create mate references**

If you are creating a part that will be used in multiple assemblies, such as a library or commercial-off-the-shelf (COTS) part, consider adding a Mate Reference (Fig.5.6) or two[3]. This will allow you to drag in these parts and quickly apply mates. And if you name the references, your part can be automatically mated when placed in the assembly.

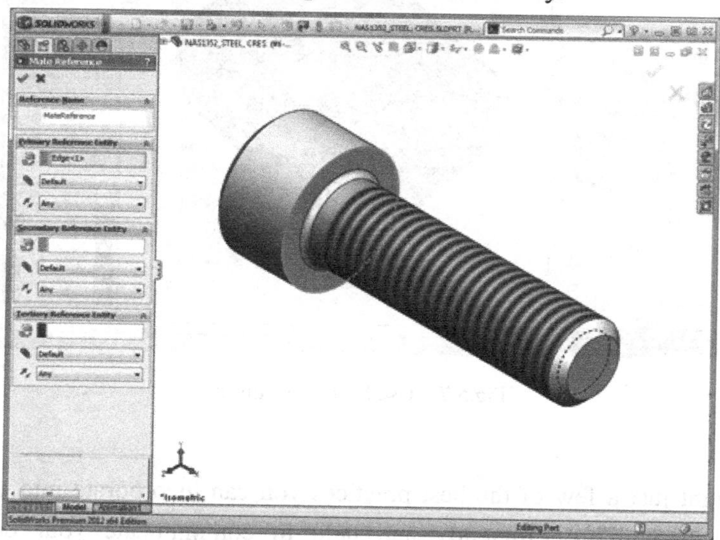

Fig.5.6 Create Mate References

★ **Apply cosmetic fillets and chamfers last**

By not adding any critical features (such as cosmetic fillets and chamfers) until the end of the feature tree, you decrease the amount of rebuild time for the part.

Adding cosmetic features last will also make it easier to create simplified configurations of your models for analysis or to speed up assembly performance, as they will be easy to find at the end of the FeatureManager tree and there will be no dependencies to other geometry that may cause issues when they are suppressed.

★ **Use fillet features instead of sketch fillets**

There are some instances where a sketch fillet makes sense. But in my experience, I have found that in most cases, a fillet feature will result in a more robust model and also makes troubleshooting easier. Using fillet features also allows you to group fillets based on the fillet radius when appropriate, and can allow the model to be simplified easier.

★ **Use library features**

If you find yourself (or your co-workers) constantly reusing the same feature(s), such as connector holes or sheet metal louvers, consider creating a Library Feature (Fig.5.7). This will allow you to drag and drop the feature onto your model[4]. You can set up the Library Feature to take user input, and you can have configurations, such as shell sizes for connectors.

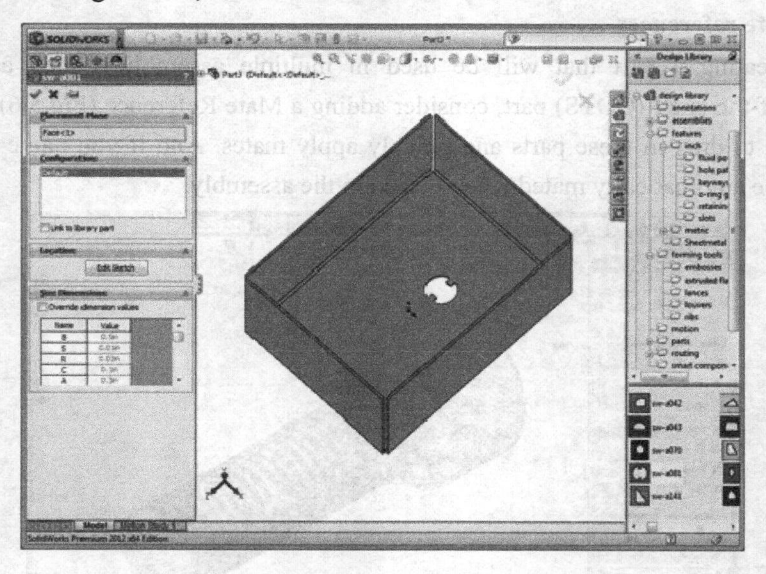

Fig.5.7 Use Library Features

★ **Conclusion**

These tips highlight just a few of the best practices you can incorporate into your workflow to help you model parts more efficiently and help to communicate your design intent. By incorporating some (or all) of these suggestions, my hope is that you will become a better and more efficient SolidWorks user.

Words

model	*n.* 模型
streamline	*n.* 流线；流线型 *vt.* 把……做成流线型；组织；使合理化；使简单化 *adj.* 流线型的
template	*n.* 模板
property	*n.* 属性
origin	*n.* 原点
symmetry	*n.* 对称
alphabetical	*adj.* 按字母顺序排列的
symmetrical	*adj.* 对称的
consistency	*n.* 一致，一致性
fraction	*n.* 分数
multiple	*n.* 倍数
cosmetic	*adj.* 修饰的，装饰的，美容的；化妆用的
robust	*adj.* 稳健的
fillet	*n.* 圆角
chamfer	*n.* 倒角
configuration	*n.* 配置
troubleshoot	*vt.* 寻找故障；故障排解

Notes

[1] By incorporating some best practices into your SolidWorks modeling routine, you can streamline your design work and focus on your design instead of driving SolidWorks.

【注释】modeling routine：常规建模。SolidWorks 中的建模主要指几何建模，是以几何信息和拓扑信息反映结构体的形状、位置、表现形式等数据的方法。

[2] You can also create conditional situations using more complex equations and VBA. For instance, if you wanted to create a dimension whose value depends on a range of values for another dimension, you can use the Visual Basic Immediate If (IIf) function.

【注释】

① Visual Basic for Applications(VBA)是 Visual Basic 的一种宏语言，是微软开发出的用于其桌面应用程序中，执行通用的自动化(OLE)任务的编程语言。它不仅是一种应用程式视觉化的 Basic 脚本，也是寄生于 VB 应用程序的版本。

② IIf 是 VBA 中条件运算指令，格式为 IIf(参数 1，参数 2，参数 3)，参数 1 是个条件判断，如果条件为真，那么返回参数 2；否则，返回参数 3。

[3] If you are creating a part that will be used in multiple assemblies, such as a library or commercial-off-the-shelf (COTS) part, consider adding a Mate Reference or two.

【注释】

①commercial-off-the-shelf (COTS)：“商用现成品或技术”或者“商用货架产品”，指可以采购到的具有开放式标准定义接口的软件或硬件产品，可以节省成本和时间。

②Mate Reference：配合参考。配合参考是一种智能配合技术，用户通过在零件中预先定义要作为配合参考的面、边线或顶点，保存零件后，便可以在装配体中插入该零件，实现智能配合。

[4] If you find yourself (or your co-workers) constantly reusing the same feature(s), such as connector holes or sheet metal louvers, consider creating a Library Feature. This will allow you to drag and drop the feature onto your model.

【注释】

① Library Feature：库特征。SolidWorks 提供了很多可以被引用的特征库类型让使用者快速地建立模型，同样也可建立特定的库特征(如键槽、中心孔等)。通过建立库特征，加快设计速度，提高效率。

② drag and drop：拖放，SolidWorks 中借助鼠标的一种操作。

Part B　Reading Materials

Viewports

There are things within SolidWorks that can help speed things up for you that aren't usually taught during training classes. These are things that are built into the software out of the box, but aren't necessarily *out in the open*.

If you need to create a tangent arc (Fig.5.8) from the line you're sketching, there are two easy ways to accomplish it without having to actually select the Tangent Arc tool. Start your line segment (click and release), draw your line, then click to start the tangent section. At this point you can either hit the 'A' key on your keyboard, or you can pull away a bit from the end point, then go back over it. Either of these will switch your tool from line to tangent arc. You'll notice a concentric relationship between the two endpoints.

Also, should you realize that you didn't want a tangent arc there, hit the 'A' key to go back to your line tool.

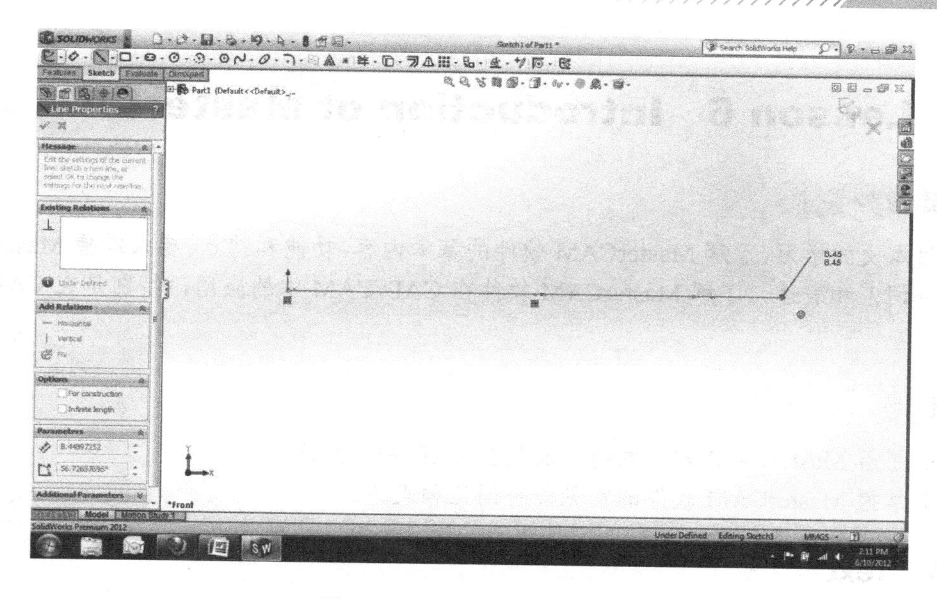

Fig.5.8 Create a Tangent Arc

Part C Exercises

Translate the following paragraphs into Chinese.

Creating a Project

Click FloWorks, Project, Wizard. The project wizard guides you through the definition of a new COSMOSFloWorks project.

In the Project Name dialog box, click Use current (40 degrees). Each COSMOSFloWorks project is associated with a SolidWorks configuration. You can attach the project either to the current SolidWorks configuration or create a new SolidWorks configuration based on the ce-rret one.

In the Units dialog box, you can select the desired system of units for both input and output (results). For this project use the International System SI by default.

In the Analysis Type dialog box, you can select either Internal or External type of the flow analysis. Specify internal type and accept the other default settings.

In the Roughness dialog box, you can specify the wall roughness value applied by default to all model walls. To specify the roughness value for a specific wall you must create a Real Wall boundary condition. In this project we will not concern rough walls.

Lesson 6 Introduction of MasterCAM

教学目的和要求

通过本文的学习，了解 MasterCAM 软件的基本内容、功能和特点。要求掌握 MasterCAM 常见命令词汇和表述，了解 MasterCAM 软件在 CAD/CAM 中的应用，特别是在 CAD 中的应用。

重点和难点

（1）了解 MasterCAM 软件的特点及其在 CAD 中的应用。

（2）掌握 MasterCAM 软件的常见命令词汇和表述。

Part A Text

Founded in MA in 1983, CNC Software, Inc. is one of the oldest developers of PC-based computer-aided design/computer-aided manufacturing (CAD/CAM) software. They are one of the first to introduce CAD/CAM software designed for both machinists and engineers. Mastercam, CNC Software's main product, started as a 2D CAM system with CAD tools that let machinists design virtual parts on a computer screen and also guided computer numerical controlled (CNC) machine tools in the manufacture of parts. Since then, MasterCAM has grown into the most widely used CAD/CAM package in the world. CNC Software, Inc. is now located in Tolland, Connecticut.

★ MasterCAM product levels

With the release of MasterCAM 2017, MasterCAM entered the 3rd interface in the product history moving to a familiar Windows Office tab format. Prior to MasterCAM 2017 was Version X that was introduced in 2005. With Version X the application became a true Windows-based application, as opposed to one ported over from DOS. It also represented a fundamental shift in the way the application was configured. MasterCAM supports many types of machines, each with a choice of levels of functionality, as well as offers optional add-ins for 4-axis and 5-axis machining.

The following list describes the MasterCAM product levels that were updated with the release of MasterCAM 2017.

Design—3D wireframe geometry creation, surface and solid modeling, dimensioning, importing and exporting of non-MasterCAM CAD files (such as AutoCAD, SolidWorks, Solid Edge, Inventor, Parasolid, etc.).

Mill Entry—Includes Design, basic 2-2 1/2 axis machining including associative contouring, zigzag and one-way pocketing, drilling operations and full toolpath verification.

Mill—Includes Mill Entry, 2-3 axis single surface machining plus limited multi-surface rough pocketing and finish parallel machining. Also, positional 4th axis rotary support. Includes Engrave, Nesting.

Mill 3D—Includes Mill plus additional support multi-surface machining up to 3-axis.
Multi-Axis add-on—5-Axis roughing, finishing, flowline multisurface, contour, depth cuts, drilling, advanced gouge checking.

Mill-Turn add-on—Includes full machine simulation and the ability to synchronize multiple code streams including pinch/balance turning. Requires Mill 3D or Mill Lathe, and a machine file.

Lathe Entry— Includes Design, Solids and 2-axis fully associative turning functionality and full tool-path verification.

Lathe— Full 2-axis turning functionality with C&Y-axis machining and part handling operations. Additional functionality gained by bundling with appropriate Mill package.

Router Entry— Includes Design basic 2-2 1/2 axis machining including associative contouring, zigzag and one-way pocketing, drilling operations and full toolpath verification.

Router—Includes Router Entry, 2-3 axis single surface machining plus limited multi-surface rough pocketing and finish parallel machining. Includes Engrave, Nesting.

The Router products are targeted to the woodworking industries but are virtually identical to the Mill line with additional of support for router specific functions such as aggregate heads, drill blocks and saws.

Art add-on—Quick 3D design, 2D outlines into 3D shapes, shape blending, conversion of 2D artwork into machinable geometry, plus exclusive fast toolpaths, rough and finish strategies, on-screen part cutting.

★ CAD Solutions

MasterCAM is the most widely used CAD/CAM software worldwide and remains the program of choice among CNC programmers. MasterCAM 2017 is the next generation of our popular program, delivering the most comprehensive toolpaths combined with robust CAD tools[1].

MasterCAM gives you a powerful and integrated foundation of shop-tested CAD tools. From wireframe and surfacing to solid modeling, MasterCAM ensures that you're ready for any job.

MasterCAM delivers:

- Full 3D live wireframe modeling.
- Powerful surface modeling and editing.
- Solid modeling.
- Broad range of translators allowing you to open any CAD file.
- Associative dimensioning, analysis, and much more

1) Design

MasterCAM's streamlined CAD engine makes design work easier than ever before. Each piece of geometry you create is live, letting you quickly modify it until it's exactly what you want. And, with traditional functions consolidated into a few simple clicks, MasterCAM simplifies the creation of even the most complex parts[2].

Surface modeling remains one of the best ways to create 3D organic shapes. From extending and splitting surfaces, to creating and editing shaded surface models, MasterCAM delivers a powerful set of surface design tools that give you straightforward control over the final detail of even the most complex job.

Key Features:

- Easy geometry modeling and editing.
- Create parametric and NURBS surfaces using loft, ruled, revolved, swept, draft, and offset creation methods.
- Create splines on surfaces using a variety of functions including surface projections, surface intersections, edge curves, slicing curves, and parting lines.
- Region Chaining explicitly defines the area for a 2D dynamic milling toolpath by simply selecting the face to be machined and face to be avoided. MasterCAM treats the selections as complete machining chains[3].
- Analyze surface curvature to identify undercuts and minimum radius.
- Quickly modify parameters such as line length and arc radius after creation.

2) Solids

Import, create, and program your solids in the same easy-to-use interface, with full access to all MasterCAM's powerful programming tools. From fast solid construction using familiar modeling techniques, to easy shelling and thin-wall creation, MasterCAM Solids delivers solid modeling with an eye for production.

Flexibility is crucial to efficient part design and programming. MasterCAM Solids lets you mix and match modeling techniques. Add surface or wireframe elements to a solid. Quickly add solid components to a complex surface model. MasterCAM Solids gives you the speed of solids, the power of surfaces, and the simplicity of wireframe. You choose the right tool for the right job.

Key Features:

- Powerful, streamlined solid modeling for practical design.
- Easy-to-use revolve, extrude, loft, and sweep commands.
- Combine solids and surfaces in the same model.
- Trim solids to a single surface or to multiple surfaces using sheet solids for maximum design potential.
- MasterCAM's Solid Feature Recognition identifies features such as fillets and holes on imported solids and adds those features to the history tree.
- Open SOLIDWORKS® and Solid Edge® files directly.

3)Integrated CAD Solutions

"I do not have to import parameterization files anymore. I program inside of SOLIDWORKS and if anything in the design changes, MasterCAM will flag these areas as dirty operations. To clean these up, I simply select the geometry and hit regenerate and Mastercam automatically regenerates the toolpaths." Billy Crump, Programmer / Machinist, Scinomix.

We're bringing the power of MasterCAM to the most widely-used modeling software with MasterCAM® for SOLIDWORKS®. Now you can program parts directly in SOLIDWORKS, using toolpaths and machining strategies used the most by shops around the world.

MasterCAM® for SOLIDWORKS® gives your offer the best possible foundation for fast and efficient milling. From general purpose methods such as optimized pocketing to highly specialized 5-axis cutting, it ensures that you're ready for any job. MasterCAM® for SOLIDWORKS® delivers:

- Gold Partner integration with SOLIDWORKS®.
- Easy pocketing, contouring and drilling.
- Feature-based 2D programming.
- Full 3D machining with automatic leftover removal.
- Powerful multi-axis cutting.
- Solid-model cut verification.

Words

manufacture	*n.* 制造
package	*n.* 包裹；套装软件(包)，程序包
toolpath	刀具路径

robust	*adj*. 稳健的，有效的
streamlined	*adj*. 合理化的，改进的
dynamic	*adj*. 动态的
feature-based	基于特征的
optimized	*adj*. 最佳化的，优化的

Notes

[1] MasterCAM is the most widely used CAD/CAM software worldwide and remains the program of choice among CNC programmers. MasterCAM 2017 is the next generation of our popular program, delivering the most comprehensive toolpaths combined with robust CAD tools.

【译文】MasterCAM 是世界上使用最广泛的 CAD/CAM 软件，也是数控编程人员的首选方案。MasterCAM 2017 是最受欢迎的新一代产品，它提供了最全面的工具路径和性能稳健的 CAD 工具。

[2] MasterCAM's streamlined CAD engine makes design work easier than ever before. Each piece of geometry you create is live, letting you quickly modify it until it's exactly what you want. And, with traditional functions consolidated into a few simple clicks, MasterCAM simplifies the creation of even the most complex parts.

【译文】MasterCAM 合理化的 CAD 引擎使设计工作比以往任何时候都变得更容易。你创建的每一个几何图形都是“即时的”，可以让你迅速做出修改，直到它完全符合你的要求。而且，由于传统的功能被整合，只须简单地单击鼠标。可以说，MasterCAM 简化了最复杂零件的创建。

[3] Region Chaining explicitly defines the area for a 2D dynamic milling toolpath by simply selecting the face to be machined and face to be avoided. MasterCAM treats the selections as complete machining chains.

【译文】通过简单地选择待加工面和避让面，区域链接可以明确地定义二维动态铣削工具路径的区域。MasterCAM 将该选择视为完整的加工链。

Part B　Reading Materials

MasterCAM is the world's most widely-used CAD/CAM software for the 22nd straight year, according to the latest analysis by CIMdata, Inc. With over 224,000 installed seats worldwide, MasterCAM has almost twice the installed base of the closest competitor.

"CNC Software is grateful to our constantly expanding and highly experienced global user base for their continued support. It's the feedback and input from the worldwide manufacturing community that has helped MasterCAM stay at the forefront of manufacturing technology and productivity today." says CNC Software President, Meghan West.

West continues, "The large community of qualified MasterCAM programmers presents a number of opportunities. As a manufacturer, if you're looking to expand, you can find experienced talent, and as a programmer, you can always find a shop that uses MasterCAM."

CIMdata also recognized MasterCAM's global support network as the largest in the CAM world. "Our MasterCAM Resellers continue to provide manufacturers with expert training, on-site support, and consulting to ensure production goals and deadlines are met," West adds.

"And with almost 25% of the educational market," West continues, "CNC Software continues to demonstrate its commitment to the future of our industry, as well as the next generation of manufacturing professionals. We continue our commitment to all levels of learning, from secondary education and community programs to technical schools and universities. We have worked with our global support network to develop programs and tools that introduce students to the world of manufacturing at every age, and that teach the programming skills necessary for meaningful employment."

Part C Exercises

Why Choose MasterCAM?

- Streamlined — A remarkably easy interface has you cutting parts quickly.
- Comprehensive — Robust CAD tool
- ls complement your CAM programming.
- Integrated — Tightly connected CAD and CAM let you easily model for manufacturing.
- Flexible — Open virtually any CAD file for additional design work and machining.
- Supported — Work with the most professional, knowledgeable reseller network in the industry.

Lesson 7 Abaqus Unified FEA

教学目的和要求

本文为介绍 Abaqus 统一有限元分析软件的英文专业文献，主要围绕 Abaqus 功能模块进行描述和介绍，从 Abaqus 应用领域开始，分别介绍了 Abaqus/CAE、Abaqus/标准求解、Abaqus/显示求解、Abaqus 多物理场技术、Abaqus 附加模块等功能模块和相关技术。通过本文的学习，读者可以了解有关 Abaqus 统一有限元分析软件的英文表达和常见功能术语的英文名称，有助于学习和应用英文版有限元分析软件；要求掌握文中所涉及的 Abaqus 统一有限元分析软件相关术语的英文名称，在实际应用中不断积累各类相关词汇的英文名称，并能对文后所附的练习进行理解和翻译。

重点和难点

（1）重点掌握 Abaqus 统一有限元分析软件相关的专业术语及其表达。
（2）了解 Abaqus 有限元分析软件的功能、常用模块和应用领域。

Part A Text

Abaqus Overview

Today, product simulation is often being performed by engineering groups using niche simulation tools from different vendors to simulate various design attributes. The use of multiple vendor software products creates inefficiencies and increases costs. SIMULIA delivers a scalable suite of unified analysis products that allow all users, regardless of their simulation expertise or domain focus, to collaborate and seamlessly share simulation data and approved methods without loss of information fidelity[1].

The Abaqus Unified FEA[2] product suite offers powerful and complete solutions for both routine and sophisticated engineering problems covering a vast spectrum of industrial applications. In the automotive industry engineering work groups are able to consider full vehicle loads, dynamic vibration, multibody systems, impact/crash, nonlinear static, thermal coupling, and acoustic-structural coupling using a common model data structure and integrated solver technology. Best-in-class companies are taking advantage of Abaqus Unified FEA to consolidate their processes and tools, reduce costs and inefficiencies, and gain a competitive advantage.

Abaqus/CAE

With Abaqus/CAE you can quickly and efficiently create, edit, monitor, diagnose, and visualize advanced Abaqus analyses[3]. The intuitive interface integrates modeling, analysis, job management, and results visualization in a consistent, easy-to-use environment that is simple to learn for new users, yet highly productive for experienced users. Abaqus/CAE supports familiar interactive computer-aided engineering concepts such as feature-based, parametric modeling, interactive and scripted operation, and GUI customization.

Users can create geometry, import CAD models for meshing, or integrate geometry-based meshes that do not have associated CAD geometry. Associative Interfaces for CATIA V5, SolidWorks, and Pro/ENGINEER enable synchronization of CAD and CAE assemblies and enable rapid model updates with no loss of user-defined analysis features.

The open customization toolset of Abaqus/CAE provides a powerful process automation solution, enabling specialists to deploy proven workflows across the engineering enterprise. Abaqus/CAE also offers comprehensive visualization options, which enable users to interpret and communicate the results of any Abaqus analysis.

Abaqus/Standard[4]

Abaqus/Standard employs solution technology ideal for static and low-speed dynamic events where highly accurate stress solutions are critically important. Examples include sealing pressure in a gasket joint, steady-state rolling of a tire, or crack propagation in a composite airplane fuselage. Within a single simulation, it is possible to analyze a model both in the time and frequency domain. For example, one may start by performing a nonlinear engine cover mounting analysis including sophisticated gasket mechanics. Following the mounting analysis, the pre-stressed natural frequencies of the cover can be extracted, or the frequency domain mechanical and acoustic response of the pre-stressed cover to engine induced vibrations can be examined. Abaqus/Standard is supported within the Abaqus/CAE modeling environment for all common pre- and postprocessing needs.

The results at any point within an Abaqus/Standard run can be used as the starting conditions for continuation in Abaqus/Explicit. Similarly, an analysis that starts in Abaqus/Explicit can be continued in Abaqus/Standard. The flexibility provided by this integration allows Abaqus/Standard to be applied to those portions of the analysis that are well-suited to an implicit solution technique, such as static, low-speed dynamic, or steady-state transport analyses while Abaqus/Explicit may be applied to those portions of the analysis where high-speed, nonlinear, transient response dominates the solution.

Abaqus/Explicit[4]

Abaqus/Explicit is a finite element analysis product that is particularly well-suited to simulate brief transient dynamic events such as consumer electronics drop testing, automotive crashworthiness, and ballistic impact. The ability of Abaqus/Explicit to effectively handle severely nonlinear behavior such as contact makes it very attractive for the simulation of many quasi-static events, such as rolling of hot metal and slow crushing of energy absorbing devices.

Abaqus/Explicit is designed for production environments, so ease of use, reliability, and efficiency are key ingredients in its architecture. Abaqus/Explicit is supported within the Abaqus/CAE modeling environment for all common pre- and postprocessing needs.

The results at any point within an Abaqus/Explicit run can be used as the starting conditions for continuation in Abaqus/Standard. Similarly, an analysis that starts in Abaqus/Standard can be continued in Abaqus/Explicit. The flexibility provided by this integration allows Abaqus/Explicit to be applied to those portions of the analysis where high-speed, nonlinear, transient response dominates the solution while Abaqus/Standard can be applied to those portions of the analysis that are well-suited to an implicit solution technique, such as static, low-speed dynamic, or steady-state transport analyses.

Abaqus Multiphysics

The Abaqus Unified FEA product suite has significant capabilities that are used to solve multiphysics problems. These capabilities, developed over many years and fully integrated as core Abaqus functionality, have been used extensively for many engineering applications on products and engineering projects in use today.

Multiphysics technology has been a part of Abaqus from the beginning. Starting with Abaqus V2 (in 1979), Abaqus/Aqua simulates hydrodynamic wave loading on flexible structures for offshore pipelines. Through the years additional multiphysics capabilities have been added, such as fluid, thermal, and electrical couplings, to name a few.

The advantage of Abaqus Multiphysics is the ease with which Multiphysics problems can be solved by the Abaqus structural FEA user. From the same model, same element library, same material data, and same load history, an Abaqus structural FEA model can easily be extended to include additional physics interaction. No additional tools, interfaces, or simulation methodology are needed.

Abaqus Add-ons

The Abaqus Unified FEA Suite enables users to develop custom applications and integrate solutions from Alliance Partners or their own in-house applications. SIMULIA offers supported Analysis Interfaces for Moldflow, MSC.ADAMS, and MADYMO. A family of crash dummy models from FTSS is available for use in crash and occupant safety simulations. A variety of customizable Extensions for Abaqus, developed by our regional offices offer efficiencies for specific industry simulation needs such as printed wiring board (PWB) modeling, composite filament winding or sheet metal forming. SIMULIA also provides fe-safe, from Safe Technology for accurate fatigue and failure predictions.

Words

overview	*n.* 概述
niche	*n.* 有利可图的缺口，商机
attributes	*n.* 属性
scalable	*adj.* 可扩展的，可伸缩的
nonlinear	*adj.* 非线性的
coupling	*n.* 耦合；结合，连接
acoustic	*adj.* 声学的；音响的；听觉的
intuitive	*adj.* 直觉的，直观的
deploy	*vt.* 配置；展开；
gasket	*n.* [机] 垫圈；[机] 衬垫；
transient	*adj.* 短暂的，瞬态，瞬时的
crashworthiness	*n.* 耐撞性
ballistic	*adj.* 弹道的；射击的
hydrodynamic	*adv.* 水力的；流体动力学的
dummy	*adj.* 虚拟的；假的
filament	*n.* 灯丝；细丝；细线；单纤维

Notes

[1] SIMULIA delivers a scalable suite of unified analysis products that allow all users, regardless of their simulation expertise or domain focus, to collaborate and seamlessly share simulation data and approved methods without loss of information fidelity.

【注释】SIMULIA：未来的分析仿真平台。达索公司并购 Abaqus 后，将 SIMULIA 作为其分析产品的新品牌，它是一个协同、开放、集成的多物理场仿真平台。真实世界的仿真是非线性的，SIMULIA 将成为模拟真实世界仿真分析工具，它支持最前沿的仿真技术和最广泛的仿真领域。SIMULIA 为真实世界的模拟提供了开放的多物理场分析平台。

SIMULIA 将同 CATIA、DELMIA 一起，帮助用户在 PLM 中，实现设计、仿真和生产的协同工作。它把分析仿真在产品开发周期的地位提升到新的高度。

[2] Abaqus Unified FEA：意思为统一有限元分析；FEA 是 Finite Element Analysis 的缩写，意思为有限元分析。

【注释】Abaqus 是一套功能强大的工程模拟有限元软件，其解决问题的范围从相对简单的线性分析到许多复杂的非线性问题。Abaqus 包括一个丰富的、可模拟任意几何形状的单元库，并且拥有各种类型的材料模型库，可以模拟典型工程材料的性能。其中，包括金属、橡胶、高分子材料、复合材料、钢筋混凝土、可压缩超弹性泡沫材料、土壤和岩石等地质材料。

作为通用的模拟工具，Abaqus 除了能解决大量结构(应力/位移)问题，还可以模拟其他工程领域的许多问题，如热传导、质量扩散、热电耦合分析、声学分析、岩土力学分析(流体渗透/应力耦合分析)及压电介质分析。达索公司于 2005 年收购了 Abaqus 软件公司。

[3] With Abaqus/CAE you can quickly and efficiently create, edit, monitor, diagnose, and visualize advanced Abaqus analyses.

【注释】Abaqus/CAE：一个全面支持求解器的图形用户界面，即人机交互前后处理模块，是 Abaqus 公司新近开发的软件运行平台，汲取了同类软件和 CAD 软件的优点，同时与 Abaqus 求解器软件紧密结合。Abaqus/CAE 采用 CAD 方式建模和可视化视窗系统，具有良好的人机交互特性。它强大的模型管理和载荷管理手段，为多任务、多工况实际工程问题的建模和仿真提供了方便。鉴于接触问题在实际工程中的普遍性，单独设置了连接(interaction)模块，可以精确地模拟实际工程中存在的多种接触问题。还采用了参数化建模方法，为实际工程结构的参数设计与优化、结构修改提供了有力工具。

[4] Abaqus/Standard Abaqus/Explicit：

【注释】Abaqus/Standard 和 Abaqus/Explicit 是 Abaqus 的两个主要分析模块。其中，Abaqus/Standard 还有两个特殊用途的附加分析模块：Abaqus/Aqua 和 Abaqus/Design。另外，Abaqus 还有分别与 ADAMS/Flex、C-MOLD 和 MoldFlow 的接口模块：Abaqus/ADAMS，

Abaqus/C-MOLD 和 Abaqus/MoldFlow。Abaqus/CAE 是完全的 Abaqus 工作环境模块，它包括 Abaqus 模型的构造、交互式提交作业、监控作业过程以及评价结果的能力。Abaqus/Viewer 是 Abaqus/CAE 的子集，它具有后处理功能。

Abaqus/Standard 是一个通用分析模块，它能够求解领域广泛的线性和非线性问题，包括静力、动力、热和电问题的响应等。

Abaqus/Explicit 是用于特殊目的分析模块，它采用显式动力有限元列式，适用于冲击和爆炸这类短暂，瞬时的动态事件，对成型加工过程中改变接触条件的这类高度非线性问题也非常有效。两个分析模块的 Abaqus/CAE 界面是一样的，两个模块的输出也是类似的，不论哪个模块都可以采用可视化图形进行后处理。

Part B Reading Materials

Realistic Simulation for Aerospace & Defense

Aerospace & Defense

Despite escalating gas and material prices, the Aerospace & Defense industry continues to grow. Over the past few years, the industry has seen major consolidation and increasing pressure to find ways to stay competitive. The entire industry—including suppliers—needs to find ways to work globally, improve processes and leverage current systems to get to market faster, lower costs, and improve quality.

The Aerospace development process strikes an intricate balance between requirements for capacity, weight, strength, efficiency, reliability, and safety. Modern flight vehicles undergo severe operating conditions, including changes in atmospheric pressure, temperature, and structural load. These vehicles also integrate complex multi-functional subsystems that must perform reliably together. No technical uncertainty can be left unchecked.

The effective use of realistic simulation is a critical component of this process. Aerospace & Defense manufacturers and suppliers use SIMULIA solutions as part of their integrated development environment to evaluate design alternatives, collaborate on projects and leverage computing resources for more efficient analysis.

Aerostructures

Aerospace structures need to survive in some of the harshest environments imaginable. Variations from extreme heat to extreme cold, debris impact from birds and hail, structural fatigue due to extreme loading cycles and many other factors all need to be considered in aerostructure design. In addition, the joining of many parts of various materials, such as composites to metals, requires a great deal of care and attention. The behavior of complex fasteners used in aerostructure design including welds, rivets, z-pins and many others make analysis of these structures even more challenging.

SIMULIA provides realistic simulation solutions that stand up too many engineering challenges. Our products can be used to perform detailed analysis of all aspects of designs. Linear capabilities for natural frequency prediction, forced response analysis and coupling to aeroelastic analysis software provide complete coverage of modal dynamic prediction requirements. Nonlinear geometric effects, such as wing deflection under extreme load, contact between parts and panel buckling can be considered in SIMULIA's aerostructures solution. In addition, dynamic events such as bird strike, hail strike and tire debris impact can be simulated from within a single software environment.

Part C Exercises

Translate the following paragraphs into Chinese

Composite materials are increasingly utilized by many industries due to their beneficial properties and the ability to tailor their response as needed. Certification of such structures requires component and sometimes full-vehicle testing, which is expensive and time-consuming and usually occurs too late in the design cycle to have any meaningful impact on the design itself.

Until recently, analysis capabilities have not been sufficiently advanced to allow simulation to be used to influence design earlier in the development cycle.

Using the appropriate advanced simulation tools can improve the design, increase the value of testing and significantly reduce the amount and scale of physical testing which is currently required. The Abaqus FEA product suite utilizes not only static, but also dynamic load case simulation such as impact, collision, BVID and bird strike. SIMULIA's product platform can evolve your composites simulation methodology to deliver more accurate results to meet regulatory and competitive demands.

Unit Two Machine Design

Lesson 8 Machine Parts

教学目的和要求

通过本文的学习，了解齿轮、V带、链传动等通用件的概念、结构及传动特点。要求掌握不同齿轮的名称和特点，V带和链条的结构、参数和组件的英文名称并能简要描述链传动的装配结构和各零件间的装配关系。

重点和难点

（1）掌握齿轮的分类及特点。
（2）掌握V带传动的各组件名称和传动原理。
（3）难点在于链传动的结构和装配关系的表述。

Part A Text

Gears

Gears are direct contact bodies, operating in pairs, that transmit motion and force from one rotating shaft to another, or from a shaft to a slide (rack), by means of successively engaging projections called teeth[1].

Tooth profiles. The contacting surfaces of gear teeth must be aligned in such a way that the drive is positive i.e., the load transmitted must not depend on frictional contact. As shown in the treatment of direct contact bodies, this requires that the common normal to the surfaces not to pass through the pivotal axis of either the driver or the follower.

As it is known as direct contact bodies, cycloidal and involute profiles provide both a positive drive and a uniform velocity ratio, i.e., conjugate action.

Basic relations. The smaller of a gear pair is called the pinion and the larger is the gear. When the pinion is on the driving shaft the pair acts as a speed reducer. When the gear drives, the pair is a speed increaser. Gears are more frequently used to reduce speed than to increase it.

If a gear having N teeth rotates at n revolutions per minute, the product N×n has the dimension *teeth per minute*. This product must be the same for both members of a mating pair if each tooth acquires a partner from the mating gear as it passes through the region of tooth engagement.

For conjugate gears of all types, the gear ratio and the speed ratio are both given by the ratio of the number of teeth on the gear to the number of teeth on the pinion. If a gear has 100 teeth and a mating pinion has 20, the ratio is 100/20=5. Thus the pinion rotates five times as fast as the gear, regardless of the speed of the gear. Their point of tangency is called the pitch point, and since it lies on the line of centers, it is the only point at which the tooth profiles have pure rolling contact. Gears on nonparallel, non-intersecting shafts also have pitch circles, but the rolling-pitch-circle concept is not valid.

Gear types are determined largely by the disposition of the shafts. This means that if a specific disposition of the shafts is required, the type of gear will more or less be fixed. On the other hand, if a required speed change demands a certain type, the shaft positions will also be fixed.

Spur gears and helical gears. A gear having tooth elements that are straight and parallel to its axis is known as a spur gear, such as shown in the Fig.8.2 (a). Spur pair can be used to connect parallel shafts only.

If an involute spur pinion were made of rubber and twisted uniformly so that the ends rotated about the axis relative to one another, the elements of the teeth, initially straight and parallel to the axis, would become helices[2]. The pinion then in effect would become a helical gear. The helical gear and rack are shown in the Fig.8.1.

Fig.8.1 Helical Gear and Rack

Bevel and worm gears. Bevel gears as shown in the Fig.8.2 (b) and worm gears as shown in the Fig.8.2 (c). In order to achieve line contact and improve the load carrying capacity of the crossed axis helical gears, the gear can be made to curve partially around the pinion, in somewhat the same way that a nut envelops a screw. The result would be a cylindrical worm and gear. Worms are also made in the shape of an hourglass, instead of cylindrical, so that they partially envelop the gear. This results in a further increase in load-carrying capacity, as shown in the Fig.8.3.

Worm gears provide the simplest means of obtaining large ratios in a single pair. They are usually less efficient than parallel-shaft gears, however, because of an additional sliding movement along the teeth.

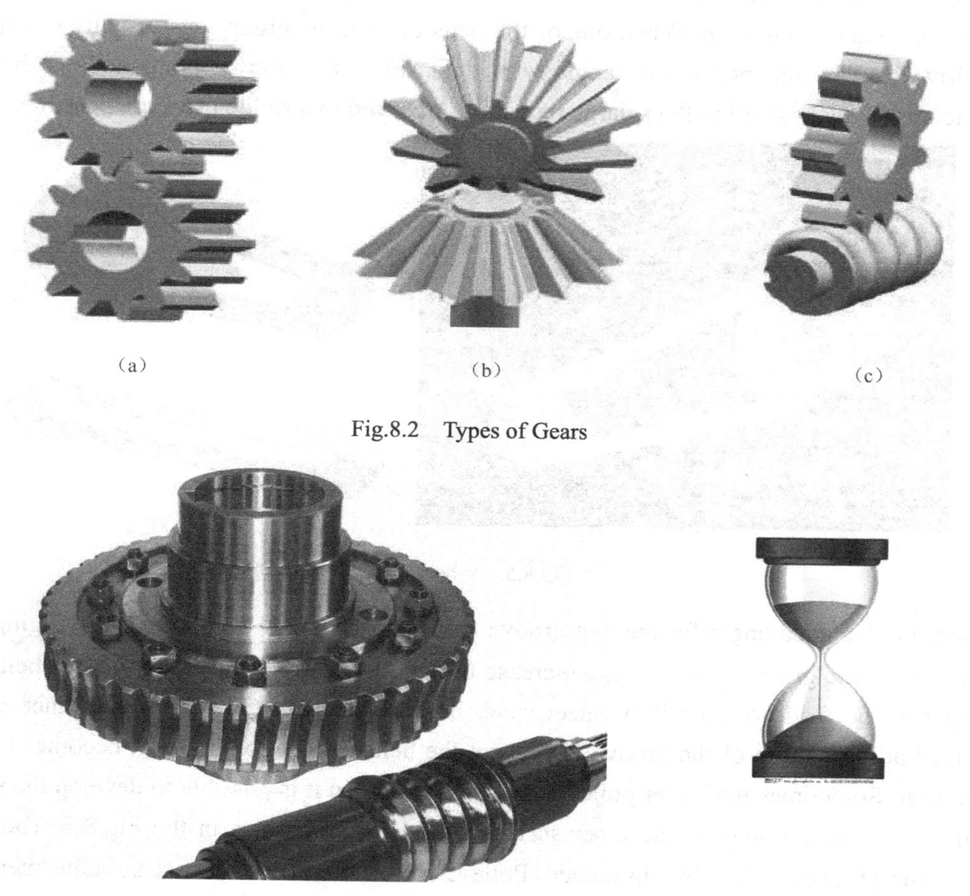

(a) (b) (c)

Fig.8.2 Types of Gears

Fig.8.3 Worm and Gear Are Increased in Load-carrying Capacity

V-belt

The rayon and rubber V-belt (Fig.8.4) are widely used for power transmission. Such belts are made in two series: the standard V-belt and the high capacity V-belt. The belts can be used with short center distances and are made endless so that difficulty with splicing devices is avoided.

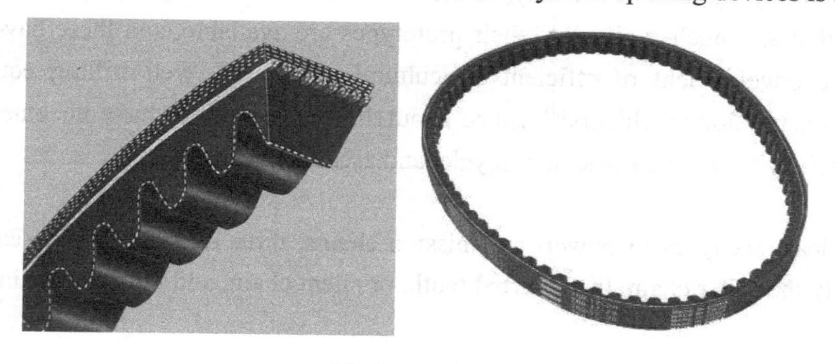

Fig.8.4 V-belt

First, cost is low, and power output may be increased by operating several belts side by side, as shown in the Fig.8.5. All belts in the drive should stretch at the same rate in order to keep the load equally divided among them. When one of the belts breaks, the group must usually be replaced. The drive may be inclined at any angle with tight side either top or bottom. Since belts can operate on relatively small pulleys, large reductions of speed in a single drive are possible.

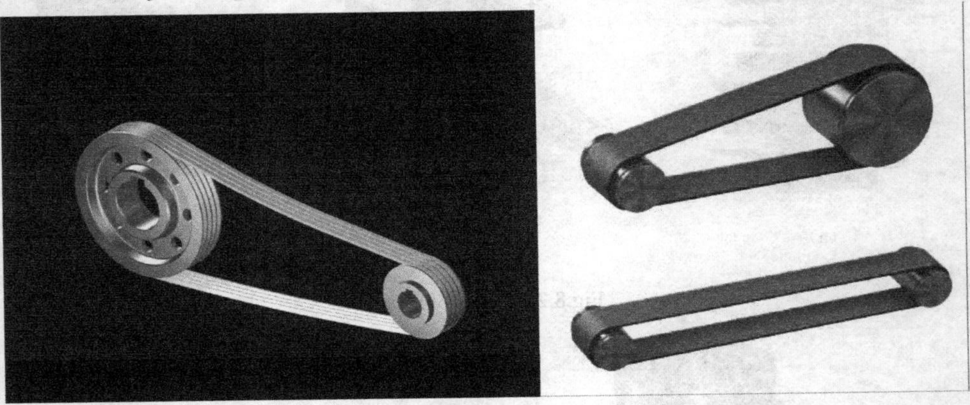

Fig.8.5 V-belt Drive

Second, the included angle for the belt groove is usually from 34° to 38° . The wedging action of the belt in the groove gives a large increase in the tractive force developed by the belt. Third, pulley may be made of cast iron, sheet steel, or die-cast metal. Sufficient clearance must be provided at the bottom of the groove to prevent the belt from bottoming as it becomes narrower from wear. Sometimes the larger pulley is not grooved when it is possible to develop the required tractive force by running on the inner surface of the belt as shown in the Fig.8.5. The cost of cutting the grooves is thereby eliminated. Pulleys are on the market permit an adjustment in the width of the groove. The effective pitch diameter of the pulley is thus varied, and moderate changes in the speed ratio can be secured.

Chain Drives

The first chain-driven or *safety* bicycle appeared in 1874, and chains were used for driving the rear wheels on early automobiles. Today, as the result of modern design and production methods, chain drives that are much superior to their prototypes are available, and these have contributed greatly to the development of efficient agricultural machinery, well-drilling equipment, and mining and construction machinery[3]. Since about 1930 chain drives have become increasingly popular, especially for power saws, motorcycle, and escalators etc.

There are at least six types of power-transmission chains, three of these will be covered in this article, namely the roller chain, the inverted tooth, or silent chain, and the bead chain, as shown in the Fig.8.6.

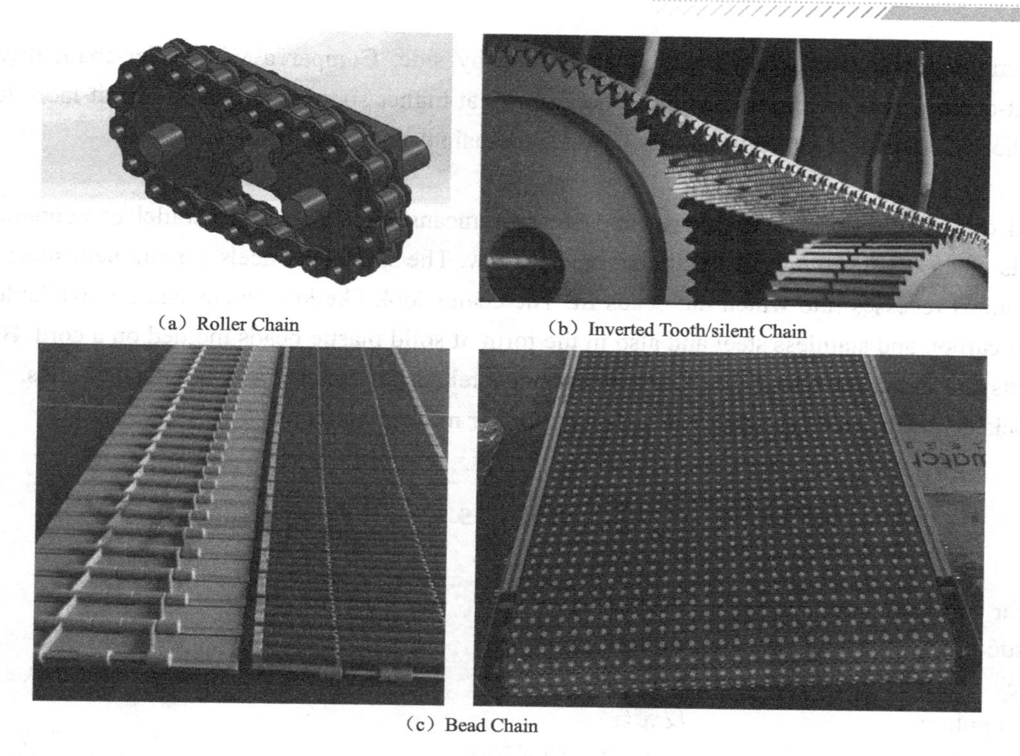
(a) Roller Chain (b) Inverted Tooth/silent Chain (c) Bead Chain

Fig.8.6 Types of the Chains

The essential elements in a roller-chain drive are a chain with side plates, pins, bushings (sleeves), and rollers, and two or more sprocket wheels with teeth that look like gear teeth. Roller chains are assembled from pin links and roller links. A pin link consists of two side plates connected by two pins inserted into holes in the side plates. The pins fit tightly into the holes, forming what is known as a press fit. A roller link consists of two side plates connected by two press-fitted bushings, on which two hardened steel rollers are free to rotate. When assembled, the pins are a free fit in the bushings and rotate slightly, relative to the bushings when the chain goes on and leaves a sprocket.

Standard roller chains are available in single strands or in multiple strands. In the latter type, two or more chains are joined by common pins that keep the rollers in the separate strands in proper alignment. The speed ratio for a single drive should be limited to about 10：1. The preferred shaft center distance is from 30 to 35 times the distance between the rollers and chain speeds greater than about 2,500 feet (800 meters) per minute are not recommended. Where several parallel shafts are to be driven without slip from a single shaft, roller chains are particularly well suited.

An inverted tooth, or silent chain is essentially an assemblage of gear racks, each with two teeth, pivotally connected to form a closed chain with the teeth on the inside, and meshing with conjugate teeth on the sprocket wheels. The links are pin-connected flat steel plates usually having straight-sided teeth with an included angle of 60 degrees. As many links are necessary to transmit the power and they are connected side by side. Compared with roller-chain drives, silent-chain drives are quieter, operate successfully at higher speeds, and can transmit more load for the same width. Some automobiles have silent-chain camshaft drives.

Bead chains provide an inexpensive and versatile means for connecting parallel or nonparallel shafts when the speed and power transmitted are low. The sprocket wheels contain hemispherical or conical recesses into which the beads fit. The chains look like key chains and are available in plain carbon and stainless steel and also in the form of solid plastic beads molded on a cord. Bead chains are used on computers, air conditioners, television tuners, and Venetian blinds. The sprockets may be steel, die-cast zinc or aluminum, or molded nylon.

Words

gear	*n*. 齿轮
slide	*n*. 滑块
rack	*n*. 齿条
belt pulley	皮带轮
projection	*n*. 凸出，凸起部分
cycloidal	*adj*. 摆线的
cycloidal profile	摆线轮廓
involute	*adj*. 渐开线的
conjugate	*adj*. 共轭的
pinion	*n*. 小齿轮
dimension	*n*. 量纲
mate	*v*. 啮合
engagement	*n*. 啮合
tangency	*n*. 相切
pitch	*n*. 节距
intersect	*v*. 相交，交叉
disposition	*n*. 排列，配置
helical gear	斜齿轮
spur gear	正齿轮，直齿轮
worm gear	蜗轮蜗杆
bevel gear	锥齿轮
hourglass	*n*. 沙漏
V-belt	*n*. V 带；三角带
groove	*n*. 沟，槽
clearance	*n*. 间隙
chain drive	链传动
sprocket	*n*. 链轮

Notes

[1] Gears are direct contact bodies, operating in pairs, that transmit motion and force from one rotating shaft to another, or from a shaft to a slide(rack), by means of successively engaging projections called teeth.

【译文】齿轮是直接接触的实体，成对使用，在称为齿的凸起的构造连续啮合的作用下，齿轮将运动和力从一根转轴传递到另一根转轴上，或者将运动和力从一根转轴传递到滑块(齿条)上。

【注释】

① operating in pairs：分词短语，修饰前面的 gears。

② that 引导的从句，修饰前面的 gears。

③ by means of 表示“借助”“通过”的意思。

[2] If an involute spur pinion were made of rubber and twisted uniformly so that the ends rotated about the axis relative to one another, the elements of the teeth, initially straight and parallel to the axis, would become helices.

【译文】假设一个渐开线直齿轮是用橡皮制成的，并且能够均匀地扭转，那么，两端就会绕着轴线作相对的转动。这样，开始时是直的且平行于轴线的小齿轮上的齿就变成了螺旋形。

【注释】

① were made of:“由……组成”。

② so that 引导结果状语从句。

③ parallel to：“平行于……”

[3] Today, as the result of modern design and production methods, chain drives that are much superior to their prototypes are available, and these have contributed greatly to the development of efficient agricultural machinery, well-drilling equipment, and mining and construction machinery.

【译文】如今，随着现代设计和制造方法的改进，链传动的应用越来越广泛，大大提高了农业机械、钻探设备、矿业和建筑机械的效率。

【注释】

① superior to 表示“优于”的意思。

② and 引导的是一句并列句。

Part B Reading Materials

A gear having tooth elements that are straight and parallel to its axis is known as a spur gear.

A spur pair can be used to connect parallel shafts only. Parallel shafts, however, can also be connected by gears of another type, and a spur gear can be mated with a gear of a different type.

Helical gears have certain advantages, for example, when connecting parallel shafts they have a higher load-carrying capacity than spur gears with the same tooth numbers and cut with the same cutter. Helical gears can be also be used to connect nonparallel, non-intersecting shafts at any angle to one another. Ninety degrees is the most common angle at which such gears are used.

Worm gears provide the simplest means of obtaining large ratios in a single pair. They are usually less efficient than parallel shaft gears, however, because of an additional sliding movement along the teeth. Because of their similarity, the efficiency of a worm and gear depends on the same factors as the efficiency of a screw.

Part C Exercises

Write the name of the following parts in the Fig.8.7 and brief description of the roller chain on the structure and assembly relations.

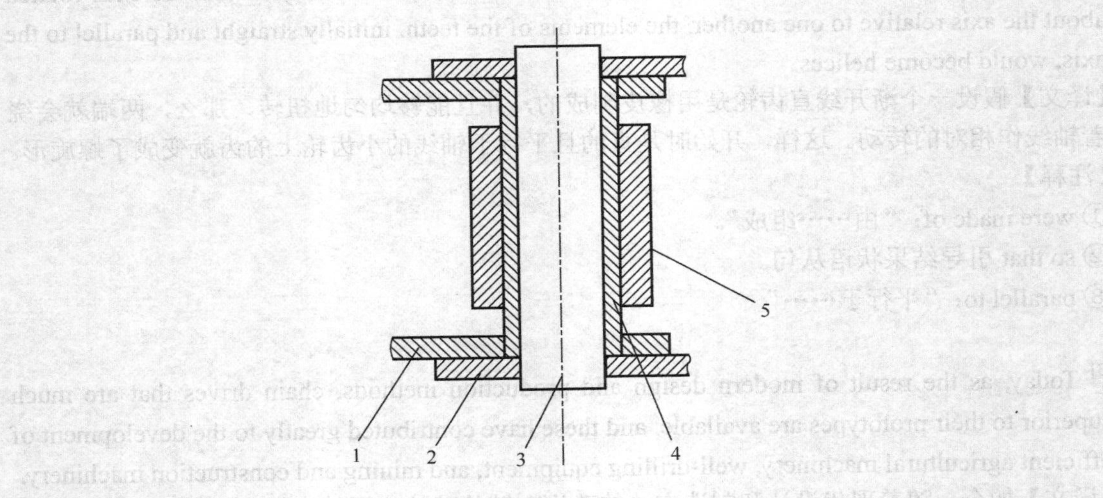

Fig.8.7 Roller Chain

Lesson 9 Mechanisms

教学目的和要求

这是一篇关于机构的说明文，通过课文的讲解，让读者了解说明文的结构和说明概念性语句的特点，结合机构的相关专业知识掌握机构的分类、专业术语和词汇的表达。

重点和难点

（1）重点掌握平面四连杆机构的专业描述和表达。

（2）掌握机构中常见的词汇和短语。

（3）对于空间机构的阅读材料，必须在充分了解空间机构的结构原理和分类的基础上才能准确理解原文和正确的翻译。

Part A Text

A mechanism is a member combination that more than two or two connections with the members realize the regulation motion made up by way of the activity. They are the component of machinery. Activity connections between two members that have the relative motion are called the motion pairs. All motion pairs contacts with planes are called lower pairs and all motion pairs contacts with points or lines are called high pairs. The motion specific property of mechanism chiefly depends on the relative size between the member, and the character of motion pairs, as well as the mutual disposition method etc[1]. The member that is used to support the member of motion in the mechanism is called the machine frame and used as the reference coordinate for the study of the motion system. The member that possesses the independence motion is called motivity member. The member except machine frame and motivity member being compelled to move in the mechanism is called driven member. The independent parameter (coordinate number) essential for description or definite mechanism motion is called the free degree of mechanism. For gaining the definite relative motion between the members of mechanism, it is necessary to make the number of motivity members of mechanism equal the number of free degrees.

Mechanisms may be categorized in several different ways to emphasize their similarities and differences. Mechanisms are generally divided into planar, spherical, and spatial categories. All three groups have many things in common, the criterion which distinguishes the groups, however, is to be found in the characteristics of the motions of the links[2].

A planar mechanism is one in which all particles describe plane curves in space and all these curves lie in parallel places ,the loci of all points are plane curves parallel to a single common plane. This characteristic makes it possible to represent the locus of any chosen point of a planar mechanism in its true size and shape on a single drawing or figure[3]. The motion transformation of any such mechanism is called coplanar. The plane four-bar linkage, the plate cam and driven parts, and the slider-crank mechanism are familiar examples of planar mechanism. The vast majority of mechanism in use today is planar.

A cam is a machine member that drives a follower through a specified motion. By the proper design of a cam, any desired motion to a machine member can be obtained. As such, cams are widely used in almost all machinery. They include internal combustion engines, a variety of machine tools, compressors and computers. In general, a cam can be designed in two ways.

(1) To design an optimal profile of a cam to give a desired motion to the follower.

(2) To choose a suitable profile to ensure a satisfactory performance by the follower.

A rotary cam is a part of a machine, which changes cylindrical motion to straight-line motion. The purpose of a cam is to transmit various kinds of motion to other parts of a machine.

Practically every cam must be designed and manufactured to fit special requirements. Though each cam appears to be quite different from the other, all of them work in similar ways. In each case, as the cam is rotated or turned, another part is connected with the cam, called a follower, is moved either right or left, up and down, or in and out. The follower is usually connected to other parts on the machine to accomplish the desired action. If the follower loses contact with the cam, it will fail to work.

According to their basic shapes, cams are classified into four different types as illustrated in Fig.9.1.

(1) Plate (Disc) Cam.

(2) Translation Cam.

(3) Cylindrical Cam.

(4) Face Cam.

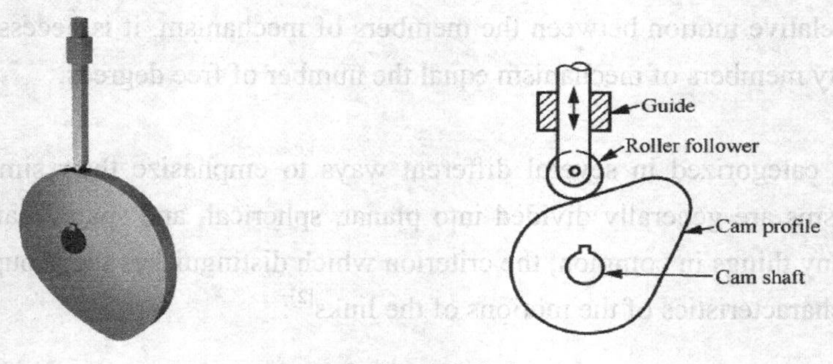

(a) Plate（Disc）Cam

Fig.9.1 Types of Cam Mechanism

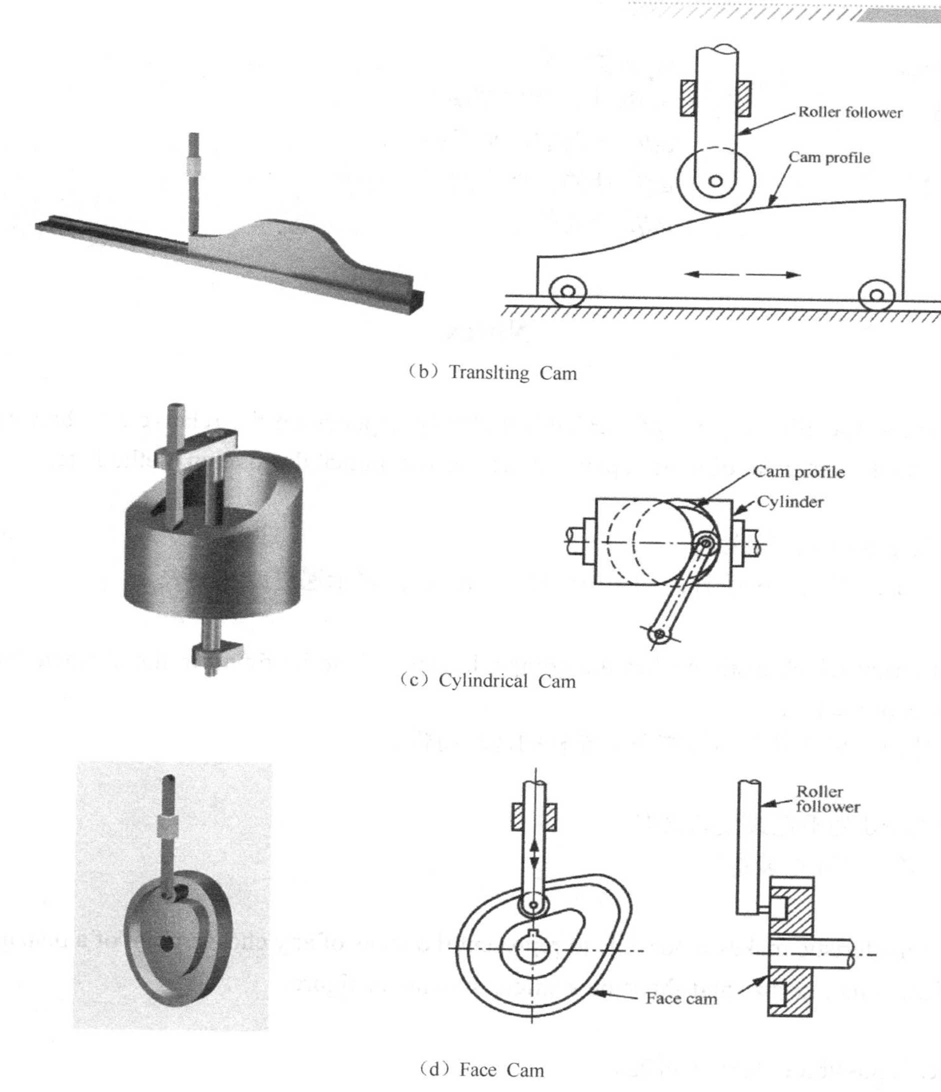

Fig.9.1 Types of Cam Mechanism(continue)

Words

motion pairs	运动副
disposition	*n.* 配置；排列
machine frame	机座，机架
coordinate	*n.* 坐标
motivity member	原动件
parameter	*n.* 参变量
driven member	从动件
free degree	自由度

categorize	*v.* 分类
category	*n.* 种类，逻辑范畴
planar	*adj.* 平面的，平坦的
spherical	*adj.* 球的，球形的
spatial	*adj.* 空间的

Notes

[1] The motion specific property of mechanism chiefly depends on the relative size between the member, and the character of motion pairs, as well as the mutual disposition method etc.

【注释】

① specific property：特性。

② as well as：不但……而且；和……一样；和；也；表示递进或并列关系。

[2] The criterion which distinguishes the groups, however, is to be found in the characteristics of the motions of the links.

【译文】然而，进行分类的标准在于连杆的运动特性。

【注释】

① to be found 为不定式被动语态。

② links 译为“连杆装置”。

[3] This characteristic makes it possible to represent the locus of any chosen point of a planar mechanism in its true size and shape on a single drawing or figure.

【注释】

① makes it possible：使……可能。

② represent：描绘，展现。

③ planar mechanism：平面机构。

④ in size and shape：在大小和形状方面。

Part B Reading Materials

A spherical mechanism is one in which each link has some point which remains stationary as the linkage moves and in which the stationary points of all links lie at a common location,

i. e. the locus of each point is a curve contained in a spherical surface, and the spherical surfaces defined by several arbitrarily chosen points axes all concentric. The motions of all particles can therefore be completely described by their radial projections, or *shadows*, on the surface of a sphere with properly chosen center.

Spherical linkages are constituted entirely of revolute pairs. A spherical pair would produce no additional constraints and would thus be equivalent to an opening in the chain, while all other tower pairs have no spherical motion. In spherical linkages, the axes of all revolute pairs must intersect at a point.

Spatial mechanisms, on the other hand, include no restrictions on the relative motions of the particles. The motion transformation is not necessary coplanar, nor must it be concentric. A spatial mechanism may have particles with loci of double curvature. Any linkage which contains a screw pair, for example, is a spatial mechanism, since the relative motion within a screw pair is helical. To take an example, worm gear pairs transmit motion between two shafts. The shafts are usually at right angles to each other but do not lie in the same plane. The teeth on the worm slide against the teeth of the worm wheel, producing a rolling action. The worm turns the worm wheel.

Part C Exercises

Translate the underlined terms and fill in the blanks according to the Fig.9.2.

1.__________ 2.__________ 3.__________ 4.__________

A.__________ B.__________

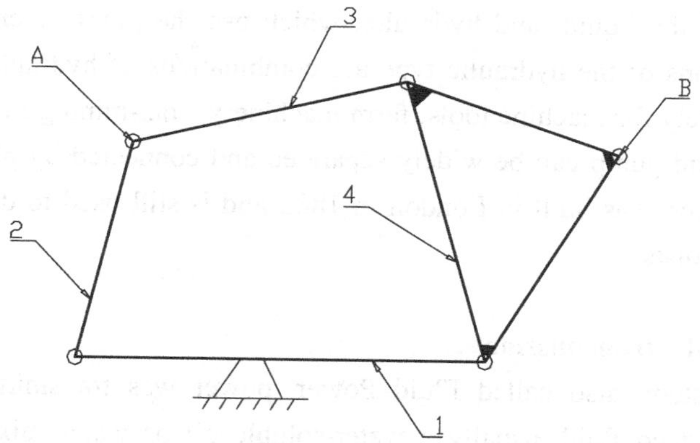

Fig.9.2 Plane Four-bar Linkage Mechanism

Lesson 10 Hydraulic Transmission

教学目的和要求

通过本文的学习，了解液压传动的概念和原理；要求掌握液压组成部分的名称及其功能，了解它们在实际生产当中的应用。

重点和难点

（1）重点掌握液压传动机构的专业描述和表达。
（2）掌握机构中常见的词汇和短语。
（3）结合液压传动的实例应用，理解液压的具体意义。

Part A Text

Introduction of Hydraulic transmission:

Hydraulic transmission, device employing a liquid to transmit and modify linear or rotary motion and linear or turning force (torque). There are two main types of hydraulic power transmission systems: hydraulic, such as the hydraulic coupling and the hydraulic torque converter, which use the kinetic energy of the liquid; and hydraulic, which use the pressure energy of the liquid[1]. Hydraulic transmissions of the hydraulic type are combinations of hydraulic pumps and motors and are used extensively for machine tools, farm machinery, coal-mining machinery, and printing presses. The motor and pump can be widely separated and connected by piping. Such a system, using pressurized water, was built in London in 1882 and is still used to drive machinery to lift bridges and operate hoists.

Tapes about Hydraulic transmission

Hydraulic power system, also called Fluid Power, power was transmitted by the controlled circulation of pressurized fluid, usually a water-soluble oil or water mixture, to a motor that converts it into a mechanical output capable of doing work on a load. Hydraulic power systems have greater flexibility than mechanical and electrical systems and can produce more power than such systems of equal size. They also provide rapid and accurate responses to controls. As a result, hydraulic power systems are extensively used in modern aircraft, automobiles, heavy industrial machinery, and many kinds of machine tools. Modern aircraft, for example, use hydraulic systems to activate their controls and to operate landing gears and brakes. Virtually all missiles, as well as their ground-support equipment, utilize fluid power. Automobiles use hydraulic-power systems in their transmissions, brakes, and steering mechanisms. Mass production and its offspring, automation, in many industries have their foundations in the utilization of fluid-power systems.

In hydraulic-power systems there are five elements: the driver, the pump, the control valves, the motor, and the load. The driver may be an electric motor or an engine of any type. The pump acts mainly to increase pressure. The motor may be a counterpart of the pump, transforming hydraulic input into mechanical output. Motors may produce either rotary or reciprocating motion in the load.

Motors in a hydraulic power system are commonly classified into two basic types: linear motors and rotational motors. A linear motor, also called a hydraulic cylinder, consists of a piston and a cylindrical outer casing. The piston constitutes the mechanical interface across which kinetic energy from the fluid is transferred to the motor mechanism. A piston rod serves to couple the mechanical force generated inside the cylinder to the external load. Hydraulic linear motors are useful for applications that require a high-force, straight-line motion and so are utilized as brake cylinders in automobiles, control actuators on aircraft, and in devices that inject molten metal into die-casting machines[2]. A rotational motor, sometimes called a rotary hydraulic motor, produces a rotary motion. In such a motor the pressurized fluid supplied by a hydraulic pump acts on the surfaces of the motor's gear teeth, vanes, or pistons and creates a force that produces a torque on the output shaft. Rotational motors are most often used in digging equipment (e.g., earth augers), printing presses, and spindle drives on machine tools.

Hydraulic cylinders (also called linear hydraulic motors) are mechanical actuators that are used to give a linear force through a linear stroke. Hydraulic cylinders are able to give pushing and pulling forces of many metric tons with only a simple hydraulic system. Very simple hydraulic cylinders are used in presses, here, the cylinder consists of a volume in a piece of iron with a plunger pushed in it and sealed with a cover. By pumping hydraulic fluid in the volume, the plunger is pushed out with a force of plunger-area pressure.

More sophisticated cylinders have a body with end cover, a piston rod, and a cylinder head. At one side the bottom is, for instance, connected to a single clevis, whereas at the other side, the piston rod is also foreseen with a single clevis. The cylinder shell normally has hydraulic connections at both sides, that is, a connection at the bottom side and a connection at the cylinder head side. If oil is pushed under the piston, the piston rod is pushed out and oil that was between the piston and the cylinder head is pushed back to the oil tank.

The hydraulic coupling is a device that links two rotation shafts. It consists of a vane impeller on the drive shaft facing a similarly vane runner on the driven shaft, both impeller and runner being enclosed in a casing containing a liquid[3]. If there is no resistance to the turning of the driven shaft, rotation of the drive shaft will cause the driven shaft to rotate at the same speed. A load applied to the driven shaft will slow it down, and a torque, or turning moment, which has the same magnitude on both shafts will be developed. In a properly designed hydraulic coupling, under normal loading conditions, the speed of the driven shaft is about 3 percent less than the speed of the drive shaft. By means of a scoop tube, the quantity of liquid in a coupling and the speed of the driven shaft can be varied. Since there is no mechanical connection between the impeller and the runner, a hydraulic coupling does not transmit shocks and vibrations.

Words

transmit	*v.* 发射，发送
torque	*n.* 扭矩
pump	*n.* 泵
fluid	*n.* 流体
automobile	*n.* 汽车
offspring	*n.* 产物，结果
vane	*n.* 叶片
pistons	*n.* 活塞
spindle	*n.* 主轴
actuators	*n.* 执行器
cylinder	*n.* 汽缸，圆筒
tank	*n.* 油箱
impeller	*n.* 叶轮
casing	*n.* 壳，管
resistance	*n.* 抵抗力，阻力
magnitude	*n.* 大小，量纲
vibration	*n.* 振动

Notes

[1] There are two main types of hydraulic power transmission systems: hydraulic, such as the hydraulic coupling and the hydraulic torque converter, which use the kinetic energy of the liquid, and hydraulic, which use the pressure energy of the liquid.

【译文】液压传动系统主要有两种类型：液力耦合器和液力变矩器，它们利用液体的动能；液力传动系统利用液体的压力能。

[2] Hydraulic linear motors are useful for applications that require a high-force, straight-line motion and so are utilized as brake cylinders in automobiles, control actuators on aircraft, and in devices that inject molten metal into die-casting machines.

【译文】液压直线电动机应用于需要高载荷的直线运动，因此可用作汽车中的制动汽缸、飞机上的控制执行器以及将熔融金属注入压铸机的设备。

[3] It consists of a vane impeller on the drive shaft facing a similarly vane runner on the driven shaft, both impeller and runner being enclosed in a casing containing a liquid.
【译文】它由驱动轴上的叶片叶轮组成，在从动轴上，面向类似的叶片流道，叶轮和流道都封闭在一个装有液体的壳体中。

Part B Reading Materials

The hydraulic torque converter is similar to the hydraulic coupling, with the addition of a stationary vaned member interposed between the runner and the impeller. All three elements are enclosed in a casing containing a liquid, usually oil. The effect of the stationary member is to make the torque, or turning moment, on the driven shaft greater than the torque on the drive shaft. When the driven shaft is stopped (stalled), the torque on it is a maximum and may be as much as 3.5 times the drive-shaft torque. A hydraulic torque converter acts like an infinitely variable speed transmission, delivering its higher torques when the output speed is low. In automatic transmissions for automobiles, it can be used as a partial or total substitute for a gearbox and clutch.

Part C Exercises

Talk about your understanding of hydraulic system application by reading the following article.

Hydraulic machines are machinery and tools that use liquid fluid power to do simple work, operated by the use of hydraulics, where a liquid is the powering medium. In heavy equipment and other types of machine, hydraulic fluid is transmitted throughout the machine to various hydraulic motors and hydraulic cylinders and becomes pressurized according to the resistance present. The fluid is controlled directly or automatically by control valves and distributed through hoses and tubes.

Hydraulic systems use liquid fluid power to do simple work, operated by the use of hydraulics, where a liquid is the powering medium. Heavy equipment is a common example. In this type of machine, hydraulic fluid is transmitted throughout the machine to various hydraulic motors and hydraulic cylinders and becomes pressurised according to the resistance present. The fluid is controlled directly or automatically by control valves and distributed through hoses and tubes.

Lesson 11 Pneumatic Transmission

教学目的和要求

通过本文的学习，了解气压传动的原理和应用，要求掌握常见的气动应用。

重点和难点

（1）了解气动力学的专有名词解释及释义。

（2）掌握气动力学的运转方式及构造。

Part A Text

Pneumatics (From Greek: πνεύμα) is a branch of engineering that makes use of gas or pressurized air. Pneumatic systems used in industry are commonly powered by compressed air or compressed inert gases[1]. A centrally located and electrically powered compressor powers cylinders, air motors, and other pneumatic devices. A pneumatic system controlled through manual or automatic solenoid valves is selected when it provides a lower cost, more flexible, or safer alternative to electric motors and actuators. Pneumatics also has applications in dentistry, construction, mining, and other areas.

★ Compressor

A mechanical device that increases the pressure of a gas by reducing its volume. An air compressor is a specific type of gas compressor.

Compressors are similar to pumps: both increase the pressure on a fluid and both can transport the fluid through a pipe. As gases are compressible, the compressor also reduces the volume of a gas. Liquids are relatively incompressible, while some can be compressed, the main action of a pump is to pressurize and transport liquids.

★ Pneumatic cylinder

Like hydraulic cylinders, something forces a piston to move in the desired direction. The piston is a disc or cylinder, and the piston rod transfers the force it develops to the object to be moved[2]. Engineers sometimes prefer to use pneumatics because they are quieter, cleaner, and do not require large amounts of space for fluid storage.

Because the operating fluid is a gas, leakage from a pneumatic cylinder will not drip out and contaminate the surroundings, making pneumatics more desirable where cleanliness is a requirement. For example, in the mechanical puppets of the Disney Tiki Room, pneumatics are used to prevent fluid from dripping onto people below the puppets.

One major issue engineer come across working with pneumatic cylinders has to do with the compressibility of a gas. Many studies have been completed on how the precision of a pneumatic cylinder can be affected as the load acting on the cylinder tries to further compress the gas used. Under a vertical load, a case where the cylinder takes on the full load, the precision of the cylinder is affected the most[3]. A study at the National Cheng Kung University in Taiwan, concluded that the accuracy is about ± 30 nm, which is still within a satisfactory range but shows that the compressibility of air has an effect on the system[4].

★ Pneumatic actuator

A Pneumatic actuator mainly consists of a piston or a diaphragm which develops the motive, power[5]. It keeps the air in the upper portion of the cylinder, allowing air pressure to force the diaphragm or piston to move the valve stem or rotate the valve control element.

Valves require little pressure to operate and usually double or triple the input force. The larger the size of the piston, the larger the output pressure can be. Having a larger piston can also be good if air supply is low, allowing the same forces with less input. These pressures are large enough to crush objects in the pipe. On 100 kPa input, you could lift a small car (upwards of 1,000 lbs.) easily, and this is only a basic, small pneumatic valve. However, the resulting forces required of the stem would be too great and cause the valve stem to fail.

This pressure is transferred to the valve stem, which is connected to either the valve plug, butterfly valve etc. Larger forces are required in high pressure or high flow pipelines to allow the valve to overcome these forces, and allow it to move the valves moving parts to control the material flowing inside.

The valves input is the "control signal." This can come from a variety of measuring devices, and each different pressure is a different set point for a valve[6]. A typical standard signal is 20～100 kPa. For example, a valve could be controlling the pressure in a vessel which has a constant out-flow, and a varied in-flow (varied by the actuator and valve). A pressure transmitter will monitor the pressure in the vessel and transmit a signal from 20 to 100 kPa. 20 kPa means there is no pressure, 100 kPa means there is full range pressure (can be varied by the transmitters calibration points). As the pressure rises in the vessel, the output of the transmitter rises, this increase in pressure is sent to the valve, which causes the valve to stroke downward, and start closing the valve, decreasing flow into the vessel, reducing the pressure in the vessel as excess pressure is evacuated through the out flow. This is called a direct acting process.

Words

inert gases	*n.* 惰性气体
dentistry	*n.* 牙科
compressor	*n.* 压缩机
piston	*n.* 活塞
leakage	*n.* 泄漏
valve stem	*n.* 阀杆

Notes

[1] Pneumatic systems used in industry are commonly powered by compressed air or compressed inert gases.

【译文】工业上使用的气动系统通常由压缩空气或压缩惰性气体提供动力。

[2] The piston is a disc or cylinder, and the piston rod transfers the force it develops to the object to be moved.

【译文】活塞是一个圆盘或圆柱体，活塞杆将它所产生的力传递给要移动的物体。

[3] Under a vertical load, a case where the cylinder takes on the full load, the precision of the cylinder is affected the most.

【译文】在垂直载荷下，当汽缸承受满负荷时，汽缸的精度受影响最大。

[4] A study at the National Cheng Kung University in Taiwan, concluded that the accuracy is about ± 30 nm, which is still within a satisfactory range but shows that the compressibility of air has an effect on the system.

【译文】中国台湾成功大学的一项研究表明，精度为±30 nm左右，仍在满意的范围内，但空气的压缩性对系统有影响。

[5] A Pneumatic actuator mainly consists of a piston or a diaphragm which develops the motive power.

【译文】气动执行器主要由活塞或膜片构成，以产生动力。

[6] For example, a valve could be controlling the pressure in a vessel which has a constant out-flow, and a varied in-flow (varied by the actuator and valve).

【译文】例如，阀门可以控制容器中的压力，该容器具有恒定的流出量和变化的流量（由制动器和阀门改变）。

Part B　Reading Materials

In the pneumatic drive system, the pneumatic drive system can be divided into the following four components according to the different functions of pneumatic components and devices.

(1) Air source device: The gas source device converts the mechanical energy provided by the original motor into the pressure energy of the gas and provides compressed air for the system. It is mainly composed of air compressor, also equipped with gas storage tank, air source purification equipment and other ancillary equipment.

(2) Actuating elements: The actuator ACTS as an energy conversion, converting the pressure energy of compressed air into the mechanical energy of the working device. Its main forms are the straight line reciprocating mechanical energy output by cylinder, the swinging cylinder and the air motor respectively output the rotary and swinging mechanical energy. For the system with vacuum pressure as the power source, the vacuum sucker is used to complete all kinds of lifting operations.

(3) Control elements: The control elements are used to throttle and control the pressure, flow and flow direction of compressed air so that the system actuator works according to the functional requirements of the program and performance. According to different functions, there are many kinds of control elements, and pressure, flow, direction and logic are generally included in the pneumatic drive system.

(4) Auxiliary elements: Auxiliary element is used in element internal lubrication, the exhaust noise, the connections between components and the signal conversion, display, amplification, detection and so on all kinds of pneumatic components, such as oil mist, muffler, pipe fittings and pipe joint, converter, displays, sensors, etc.

Part C Exercises

Translate the underlined terms and fill in the blanks according to the Fig.11.1.

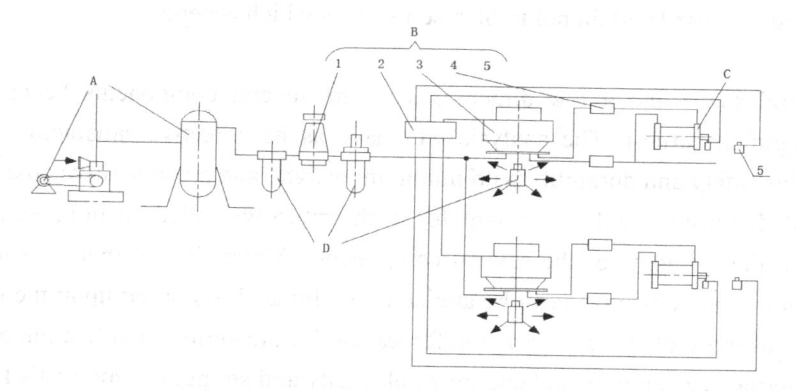

Fig.11.1 Pneumatic Transmission System

A ____________ B ____________ C ____________ D ____________

Lesson 12 Mechanical Design

教学目的和要求

本文为介绍机械设计的英文专业文献，主要围绕机械设计过程、机械设计中涉及的专业知识以及设计人员的素质等进行描述和介绍。通过本文的学习，读者可以了解有关机械设计中常见专业词汇的英文表达和专业术语的英文名称，有助于全面了解机械设计过程所涉及的专业知识；要求掌握文中所涉及的机械设计相关术语、相关学科的英文表达，在实际应用中不断积累各类专业术语的英文名称，并且能对文后所附的练习进行翻译。

重点和难点

（1）重点掌握机械设计相关的专业术语及其英文表达。

（2）了解机械设计过程和相关方法，能阅读与机械设计有关的通用英文文献。

Part A Text

The Design Process

Designing starts with a need, real or imagined. Existing apparatus may need improvements in durability, efficiency, weight, speed, or cost. New apparatus may be needed to perform a function previously done by men, such as computation, assembly, or servicing.

In the design preliminary stage, the design personnel should be allowed to fully display the creativity, not each kind of restraint. Even if has had many impractical ideas, also can in the design early time, namely in front of the plan blueprint is corrected[1]. Only then, can the mentality not be sent to stop up the innovation. Usually, the design personnel must propose several sets of design proposals, then perform the comparison. Has the possibility very much in the plan which finally designated, has used certain not in plan some ideas which accepts.

When the general shape and a few dimensions of the several components become apparent, analysis can begin in earnest. The analysis will have as its objective satisfactory or superior performance, plus safety and durability with minimum weight, and a competitive cost[2]. Optimum proportions and dimensions will be sought for each critically loaded section, together with a balance between the strengths of the several components. Materials and their treatment will be chosen. These important objectives can be attained only by analysis based upon the principles of mechanics, such as those of static for reaction forces and for the optimum utilization of friction, of dynamics for inertia, acceleration, and energy, of elasticity and strength of materials for stress and deflection, and of fluid mechanics for lubrication and hydrodynamic drives[3]. The analyses may be made by the same engineer who conceived the arrangement of mechanisms, or, in a large company, they may be made by a separate analysis division or research group. Design is a reiterative and cooperative process, whether done formally or informally, and the analyst can contribute to phases other than his own. Product design requires much research and development. Many concepts of an idea must be studied, tried, and then either used or discarded. Although the content of each engineering problem is unique, the designers follow the similar process to solve the problems.

Machine Design

The complete design of a machine is a complex process. The machine design is a creative work. Project engineer not only must have the creativity in the work, but also must have the deep elementary knowledge in aspect such as mechanical drawing, kinematics, engineering material, materials mechanics and machine manufacture technology and so on.

One of the first steps in the design of any product is to select the material from which each part is to be made. Numerous materials are available to today's designers. The function of the product, its appearance, the cost of the material, and the cost of fabrication are important in making a selection. A careful evaluation of the properties of a material must be made prior to any calculations.

Careful calculations are necessary to ensure the validity of a design. In case of any part failures, it is desirable to know what was done in originally designing the defective components. The checking of calculations (and drawing dimensions) is of utmost importance. The misplacement of one decimal point can ruin an otherwise acceptable project. All aspects of design work should be checked and rechecked.

The computer is a tool helpful to mechanical designers to lighten tedious calculations, and provide extended analysis of available data. Interactive systems, based on computer capabilities, have made possible the concepts of computer aided design (CAD) and computer-aided manufacturing (CAM)[4]. How the psychologist frequently discusses causes the machine which the people adapt them to operate. Designs personnel's basic responsibility is to diligently cause the machine to adapt the people. This certainly is not an easy work, because it is not necessary to all people to say in fact all is the most superior operating area and the operating process. Another important question, project engineer must be able to carry on the exchange and the consultation with other concerned personnel[5]. In the initial stage, the design personnel have to carry on the exchange and the consultation on the preliminary design with the administrative personnel, and is approved. This generally is carried on through the oral discussion, the schematic diagram and the writing material.

If front sues, the machine design goal is the production can meet the human need of the product.

The invention, the discovery and technical knowledge itself certainly not necessarily can bring the advantage to the humanity, only has the benefit when they are applied to produce on the product. Thus, we should realize to carries on before the design in a specific product, must first determine whether the people do need this kind of product. We must regard the machine design as a good opportunity the machine design personnel carries on the product design using creative ability, the system analysis and a formulation product manufacture technology. To grasps the project elementary knowledge is more important than to have to memorize some data and the formula. The merely service data and the formula is insufficient to the completely decision which makes in a good design need. On the other hand, a design personnel should be earnest precisely carries on all operations. For example, even if places wrong a decimal point position, also can cause the correct design to turn wrongly[6].

Good design personnel should dare to propose the new idea, and moreover is willing to undertake the certain risk. When the new method is not suitable, original method can be used. Therefore, a designer has to have the patience, because spending the time and the endeavor certainly cannot guarantee brings successfully. A brand-new design requests the screen to abandon many obsolete methods known very well for the people. Because many people of conservativeness, this is certainly not an easy matter. A mechanical designer should unceasingly explore the method of improving the existing product, should earnestly choose originally, the process confirmation principle of design in this process, with has not unified it after the confirmation new idea.

Words

durability	*n*. 耐久性
mentality	*n*. 心态；[心理] 智力；精神力
earnest	*n*. 认真；诚挚
treatment	*n*. 治疗，疗法；处理
inertia	*n*. 惯性
elasticity	*n*. 弹性，弹性（物理学），弹力
deflection	*n*. 偏向；挠曲；偏差
lubrication	*n*. 润滑
hydrodynamic	*adj*. 水力的；流体动力学的
reiterative	*adj*. 反复的；迭代的
kinematics	*n*. 运动学；动力学
fabrication	*n*. 制造，建造；装配；伪造物
defective	*adj*. 有缺陷的；不完美的
preliminary	*adj*. 初步的；开始的；预备的
Sue	*vt*. 控告；请求 *vi*. 控告；提出请求

patience	*n.* 耐心
brand-new	*adj.* 崭新的；最近获得的
unceasingly	*adv.* 不断地；继续地

Notes

[1] Even if has had many impractical ideas, also can in the design early time, namely in front of the plan blueprint is corrected.
【注释】该句为省略句，省略了主语“the design personnel”。
blueprint：蓝图，即设计图样。

[2] The analysis will have as its objective satisfactory or superior performance, plus safety and durability with minimum weight, and a competitive cost.
【注释】该句是句式平衡的一种表达，应为“The analysis will have satisfactory or superior performance, plus safety and durability with minimum weight, and a competitive cost as its objective”。

[3] These important objectives can be attained only by analysis based upon the principles of mechanics, such as those of static for reaction forces and for the optimum utilization of friction of dynamics for inertia, acceleration, and energy of elasticity and strength of materials for stress and deflection and of fluid mechanics for lubrication and hydrodynamic drives.
【注释】省略句，“of dynamics for inertia, acceleration, and energy of elasticity and strength of materials for stress and deflection, and of fluid mechanics for lubrication and hydrodynamic drives”前省略了“those”。

[4] Interactive systems, based on computer capabilities, have made possible the concepts of computer aided design (CAD) and computer-aided manufacturing (CAM).
【注释】该句是句式平衡的一种表达，应为“Interactive systems, based on computer capabilities, have made the concepts of computer aided design (CAD) and computer-aided manufacturing (CAM) possible.”。

[5] Another important question, project engineer must be able to carry on the exchange and the consultation with other concerned personnel.
【注释】该句是省略句，在“Another important question”之后省略了系动词“is”。

[6] For example, even if places wrong a decimal point position, also can cause the correct design to turn wrongly.
【注释】该句是省略句，省略主语“the design personnel”。

Part B Reading Materials

Uptrend and Prospective of Green Packaging Design

★ Build up conception and regulations

The 21st century is a century of green. As world economy develops, green thought has been expanded to the whole world, and has facilitated the governments to regulate legislation and to push forward the green authentication of green products. In the past, the traditional packaging design technology is focused on the direction of production in large quantity. Nowadays, green issue has been gradually taken into serious consideration internationally, which indicates that the future trend of green packaging design is taken good notice. We should keep encouraging consumers to build up the senses of green consumption and green value, purchase green products as the priority choice to gain the profit of environmental protection.

★ Uptrend and prospective

The worldwide green environment issue is required by times. As a member of global village, we are supposed to understand the trend of the times. The government, enterprise, designers and consumers should cooperate to each other to coexist with the Earth. In designing packaging, designers should consider the production and consumption of packaging, and take full consideration of design logic, design methodology and even future sales rather than just simply implementing design work. Another responsibility of packaging designer is to educate enterprise for adjusting business philosophy, establishing green sense of value, synchronizing the world by adopting the theory and action of green marketing, transmitting green information with consumers and satisfying visionary needs of consumers. The ultimate prospective of green packaging design is able to be achieved by setting the design goal as less quantity and higher quality and improving international competitive power of the industry. This study puts more emphasis on packaging designers and consumers' present cognition on green packaging design and its future development trend, but less emphasis on the related issues of green packaging design, such as life cycle of green packaging products, corporation social responsibility, marketing & sales and implementation case study, packaging waste logistics, which are suggested for future research directions.

Part C Exercises

Translate the following paragraphs into Chinese

It is perceived by packaging designers that the definition of green packaging design is based on the conception of environment protection to use paper to design packaging, based on the fundamental design conception of 5R principle which is to use simple, pure, non-polluted, recyclable and regenerative green packaging and based on the purpose of encouraging consumer to take sustainable consumption actions. The use of pure and adequate packaging material and nontoxic oil ink reduce the use of printing and adhesive glue is the design standard. In designers' conception, packaging design should be integrated with product and be enabled to have the function of reuse. Furthermore, it is needed to put emphasis on sense of vision and touching of packaging material so that the designers are able to design products that bring pleasant and close feeling to consumers and create simple and nature to green packaging according to different product attributes.

It is responded by designers in interview that it is necessary for designers to be familiar with and well-acquainted to the integrated system procedure of green packaging design because of the professional feature of packaging design. Only by being professional, the designers are able to have more comprehensive consideration to different phases of packaging design system. In planning packaging design work, as the pioneers of leading green design trend, the designers are supposed to have innovative thought, to make attempts and to discern the development uptrend of green packaging design. Designers should also follow 5R Principle to design green packaging and take priority of designing integrated packaging structure, to set the goal for less quantity and higher quality and put the concept of non-packed product in practice but also give consideration to both the innovative product technology and product marketing.

Unit Three Mechanical and Electrical Technology

Lesson 13 Mechatronic Engineering

教学目的和要求

通过本文的学习，了解机械电子工程的含义、组成及主要应用。要求掌握机械电子工程的英文专业术语和表达习惯，了解机械电子工程师的职责和所需具备的知识技能。

重点和难点

（1）重点掌握常用机械电子产品的英文术语。
（2）理解机械电子工程的含义及相关技术的专业术语表达。

Part A Text

The emergence of Mechatronic Engineering

The synergistic combination of M (Mechanical), E (Electrical/Electronic), and C (Computer) technologies eventually led to processes and products that could not even be contemplated before while attaining unprecedented levels of performance[1] (sees Table 1.1 and Table 1.2).

Table 1.1 Some Types of Mechatronic Products

TYPE	EXAMPLES
Transducers and measuring instrumentation	Ultrasonic receiver,Electronic scale
Processing machines	Turning and machining centers , Bonding machines , Robots
Industrial handlers	Robots, Component insertion machines
Drive mechanisms	CD players , Printers , Disk drives
Interface devices	Keyboards

Table1.2 Benefits of Mechatronic Engineering

BENEFIT	EXAMPLES
Faster response time	Servo-motion controller, Camera
Better wear and tear characteristics	Electronic ignition
Miniaturization potential	Camcorder
Easier maintenance and spare part replacement	Washing machine

Continue

BENEFIT	EXAMPLES
Memory and intelligence capabilities	Programmable sequence controllers
Shortened set up time	Computer numerical control (CNC) machines
Data processing and automation	CNC machines
User friendliness	Photocopier
Enhanced accuracy	Electronic calipers

Many modern products use embedded computers (computer-on-a-chip) to provide hitherto unattained functionalities exclusively through mechanical means. One may also exploit the ability of a computer to be programmed at will to add new functionalities. For instance, we can create *smart* products by programming the computer on the basis of fuzzy logic, or by making the computer behave like an artificial neural net (ANN). Thus computer technology offers an opportunity for endless product innovation. Mechatronic engineering is the emerging discipline that supports the development of this class of technological processes and products.

What is Mechatronic Engineering? What it is not?

Since mechatronic engineering is an emerging discipline, it is not surprising that its definition is still under development[2]. Among the more popular definitions is the one composed by the Industrial Research and Development Advisory Committee of the European Community: Mechatronics is *a synergistic combination of precision mechanical engineering, electronic control and systems thinking in the design of products and manufacturing processes.*

While the above definition seems to be acceptable in the short term, several simplistic views or, even, misconceptions continue to prevail. Two examples are:

◆ Mechatronics is *the application of microelectronics in mechanical engineering* (the original definition suggested by MITI of Japan)[3].

◆ Mechatronics is *a combination of mechanical engineering, electronic control and systems engineering in the design of products and processes.*

While these views are acceptable from a limited viewpoint, it is useful to clarify and/or elaborate upon them. Mechatronics does not warrant recognition as a distinct discipline if it were to be viewed merely as a summation, union or intersection of mechanical, electronic and computer principles (see upper part of Fig.13.1). The lower part of the figure illustrates a more useful view.

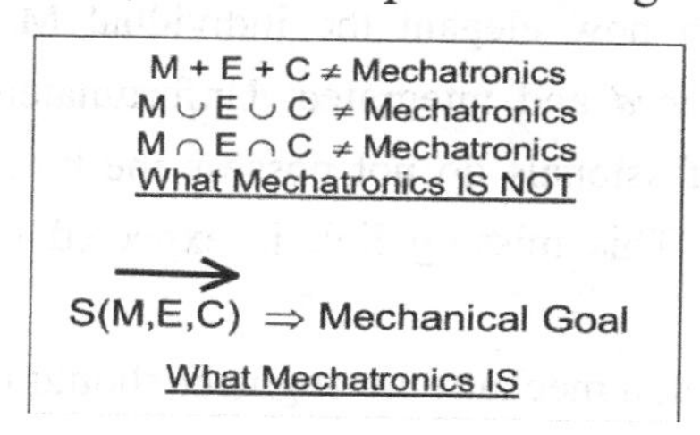

Fig.13.1 What mechatronic engineering IS and what it IS NOT

Three essential features of this view are worthy of note:

(1) There must always be a design goal that is mechanical in nature. Hence, designing a voltmeter is not a mechatronic activity although the casing of the voltmeter is mechanical in nature. The mechatronic activity needed is dictated by this *mechanical* goal. This goal is often expressed as a set of performance variables that need to be controlled (constrained within limits or optimized). Hence control engineering (especially, motion control) is central to mechatronic engineering.

(2) The design solution is invariably a system. A system is a set of interacting elements satisfying a specified goal. In the case of a mechatronic solution, the elements can be of mechanical, electronic(including electrical elements and computational elements), or software types. The interactions(signals passed) can be of analog (continuous) or digital (intermittent).

(3) Mechatronics is NOT about finding ANY solution to the given problem. Its aim is to produce a *competitive* solution. Typically, there exist a range of solutions to a given design problem. Often, a purely mechanical solution is feasible. If this is the *best* solution, it does not fall within the domain of mechatronics. Mechatronic design invariably involves tradeoffs between the advantages of alternative mechanical, electronic, and software solutions at the sub-unit level. Experience shows that an M-solution is usually inferior to a competing E-solution which in turn, is inferior to a C-solution. Hence, the hallmark of mechatronic engineering is the conscious effort to progressively substitute M-solutions by E-solutions and, in turn, E-solutions by C-solutions. This notion is signified by the *arrow* capping (M, E, C) in the lower part of Fig.13.1.

What Professional Roles Do Mechatronic Engineers Fulfill and What Knowledge and Skills Do they Need?

Technological process or product design in a modern context is usually a group activity that involves the communications among electrical, electronic, materials, mechanical, manufacturing, and other types of scientists and engineers. Usually, the team members are specialists in their respective disciplines/professions. Conventionally, such teams have been coordinated by one of the specialists with broad experience. This suffices as long the design problem is conventional. However, if the design is to be innovative, the design team needs to explore totally new avenues that involve tradeoffs amongst competing M, E, and C solutions. The competitive strength of the team then lies not so much on how elegant the individual M/E/C solutions are but in how elegantly they have been *balanced* and integrated. Unfortunately, by virtue of their specialist training, many M, E, or C professionals do not possess the breadth of knowledge required for performing this *balancing* act. This missing link is expected to be provided by mechatronic engineering professionals.

(1) From the industrial viewpoint, a mechatronic engineer should be particularly useful in Product mechatronics, i.e., the design of mechatronic products.

(2) Process mechatronics, i.e., the utilization, and maintenance of mechatronics process equipment-mainly in the manufacturing industry.

Words

mechatronic engineering	机械电子工程，机电工程
sensor	*n.* 传感器
processing machine	加工机械
bonding machine	焊接机
mechanism	*n.* 机制，机构
interface	*n.* 接口
servo	*n.* 伺服，伺服机构
wear and tear characteristics	磨损特性
maintenance	*n.* 维护，维修
spare part	备件
caliper	*n.* 卡尺
computer-on-a-chip	单片机
fuzzy logic	模糊逻辑
artificial neural net (ANN)	人工神经网络
mechatronics	*n.* 机电一体化；机械电子学
voltmeter	*n.* 电压计
actuator	*n.* 执行机构

Notes

[1] The synergistic combination of M (mechanical), E (electrical/electronic), and C (computer) technologies eventually led to processes and products that could not even be contemplated before while attaining unprecedented levels of performance.

【译文】M (机械)、E (电气/电子)和 C (计算机)技术的协同作用，最终产生了未曾料想的、具有前所未有性能或技术水平的加工过程和产品。

[2] Since mechatronic engineering is an emerging discipline, it is not surprising that its definition is still under development.

【译文】由于机械电子工程是一门新兴学科，因此，它的定义仍在发展中也就不足为奇了。

[3] Mechatronics is "*the application of microelectronics in mechanical engineering*" (the original definition suggested by MITI of Japan).

【译文】机电一体化是“微电子在机械工程中的应用”(日本通产省最初的定义)。
【注释】MITI：Ministry of International Trade and Industry 国际贸易与工业部；日本通产省

Part B　Reading Materials

Mechatronic Engineering

Mechatronic engineering is strongly based on mechanical engineering, but is a distinctly different discipline.

Many mechatronic engineers work with the electronic and computer control systems which nearly all machinery relies on for efficient and reliable operation. We take it for granted that automatic systems monitor process plants for leaks and faults, and keep the plant operating all the year round. Mechatronic engineers build and design these systems and need expertise in computing and electronics, core mechanical engineering knowledge, and the ability to bring these together to make working systems which meet the safety and reliability levels we take for granted.

Mechatronic engineers also have roles in project engineering where their crossdisciplinary knowledge gives them an edge on mechanical or electrical engineers. Mechatronic engineers can work with electrical and mechanical systems together and solve problems that cross discipline boundaries. Their strength in IT, computer hardware and networking as well as software also helps them to be very versatile problem solvers.

Mechatronic engineers also learn to develop strong team skills. At several universities, including UWA, students develop team work skills through formal instruction and self-reflection during student team projects.

Leading mechatronic engineering academics across Australia agreed on the following more formal definition for future revisions of the Engineers Australia competency standards:
Mechatronic engineering is the engineering discipline concerned with the research, design, implementation and maintenance of intelligent engineered products and processes enabled by the integration of mechanical, electronic, computer, and software engineering technologies. Specific expertise areas include:

- Artificial Intelligence Techniques
- Avionics
- Computer Hardware and Systems
- Control Systems
- Data Communications and Networks
- Dynamics of Machines and Mechanisms
- Electromagnetic Energy Conversion
- Electronics Embedded & Real-time Systems

- Fluid Power and other Actuation Devices
- Human-Machine Interface Engineering and Ergonomics
- Industrial Automation
- Measurement, Instrumentation and Sensors
- Mechanical Design and Material Selection
- Mechatronic Design and System Integration
- Modelling and Simulation
- Motion Control
- Power Electronics
- Process Management, Scheduling, Optimization, and Control
- Process Plant and Manufacturing Systems
- Robotics
- Signal Processing
- Smart Infrastructure
- Software Engineering
- Systems Engineering
- Thermofluids

Other areas of specific expertise relevant to the practice of Mechatronic engineering are found within the disciplines of Aeronautical, Engineering, Biomedical Engineering, Communications Engineering, Computer Systems Engineering, Electrical Engineering, Electrical Power Engineering, Electronic engineering, Industrial Engineering, Instrumentation and Control Engineering, Manufacturing and Production Engineering, Mechanical Engineering, Software Engineering and Space Engineering.

There are fewer job vacancies labelled “mechatronic engineer” than other disciplines. There are still not many experienced mechatronic engineers available, so most employers would not want to restrict the field of applicants by calling only for a mechatronic engineer. Mechatronic engineering positions are often advertised as:

- Asset Management engineer
- Automation engineer
- Data Logging engineer
- Electrical/Electronic engineer
- Electro mechanical engineer
- Instrumentation engineer
- Maintenance engineer
- Plant engineer
- Process engineer
- Process monitoring and plant systems engineer

- Project engineer
- Software engineer
- Systems engineer

Part C Exercises

Translate the following Fig.13.2 into Chinese.

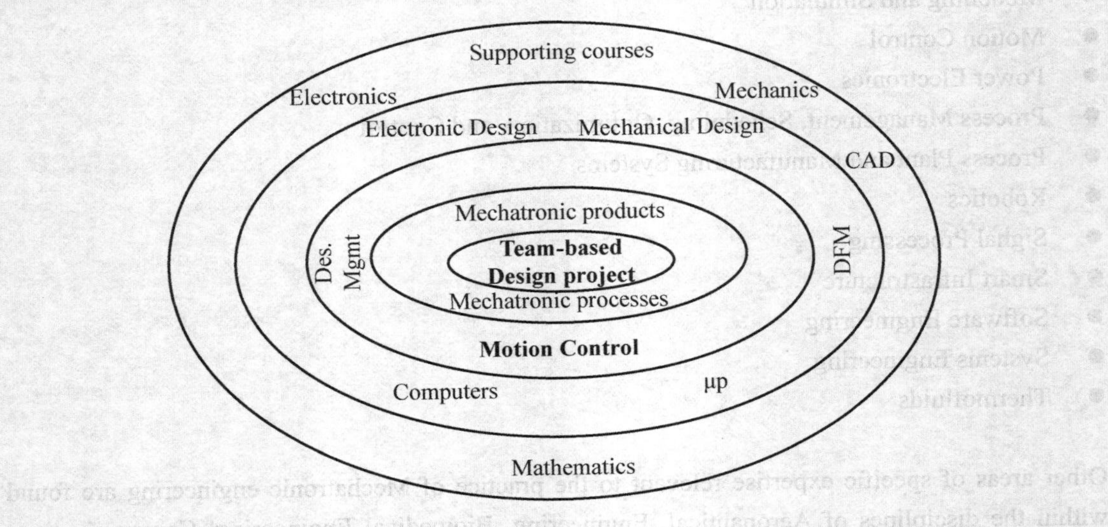

Fig.13.2 BEng. Mechatronic Eng. at City U of Hong Kong (2000)

Lesson 14 MCU Control Application

教学目的和要求

通过本文的学习，了解单片机控制系统的组成及主要应用。要求掌握单片机应用系统的英文专业术语和表达习惯。

重点和难点

（1）重点要求掌握单片机控制系统常用外围元件的名称。
（2）正确了解单片机控制系统运行过程的英文表达。

Part A Text

Four Quadrant Speed Control of DC Motor with the Help of AT89S52 Microcontroller

Introduction

DC machines play a very important role in industries and in our daily life. The outstanding advantage of DC machines is that they offer easily controllable characteristics. This paper is designed to develop a four-quadrant speed control system for a DC motor using microcontroller. The motor is operated in four quadrants i.e. clockwise, counter clock-wise, forward brake and reverse brake. It also has a feature of speed control. The four-quadrant operation of the dc motor is best suited for industries where motors are used and as per requirement they can rotate in clockwise, counter-clockwise and also apply brakes immediately in both the directions. In case of a specific operation in industrial environment, the motor needs to be stopped immediately. In such scenario, this proposed system is very apt as forward brake and reverse brake are its integral features. In this work the concept of four quadrant speed control i.e. clockwise movement, anticlockwise movement, instantaneous forward braking and instantaneous reverse braking of a dc motor with the help of microcontroller through motor driver (L293D) has been proposed[1].

Methodology

The project work has been divided into two parts. In the first part simulation is done using proteus software and in second part a prototype model is developed and the result is verified using a prototype hardware model.

1) System Overview

The design was broken down into different modules to simplify the circuit design. Fig.14.1 describes the block diagram of overall system for the four-quadrant speed control of dc motor.

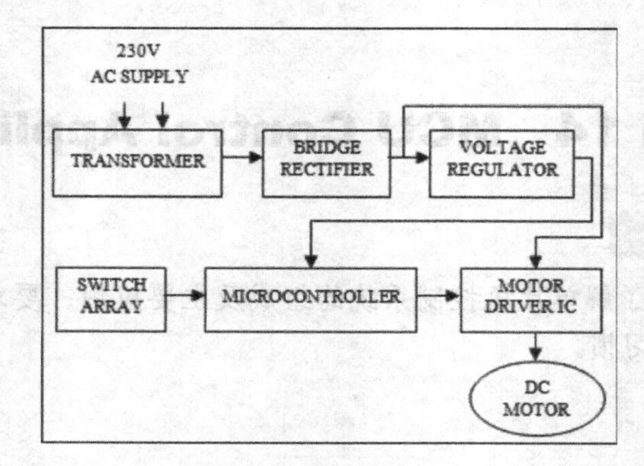

Fig.14.1 Block Diagram of the System

The circuit uses standard power supply comprising of a step-down transformer from 230V to 12V and the four diodes forming a bridge rectifier that delivers pulsating dc which is unregulated is regulated to constant 5V DC[2]. The output of the power supply which is 5V is connected to the 40pin of microcontroller and ground is connected to 20pin. Pin no 1 to 7 of port 1 are connected to switches. Pin no 21, 22, 23 of microcontroller are connected to input 1,2, enable pins of motor driver L293D.Pin 3 and 6 are connected to motor terminals.

2) Four Quadrant Operation of DC Motor

There are four possible modes or quadrants of operation using a DC Motor which is depicted in Fig.14.2. When DC motor is operating in the first and third quadrant, the supplied voltage is greater than the back emf which is forward motoring and reverse motoring modes respectively, but the direction of current flow differs. When the motor operates in the second and fourth quadrant the value of the back emf generated by the motor should be greater than the supplied voltage which are the forward braking and reverse braking modes of operation respectively, here again the direction of current flow is reversed.

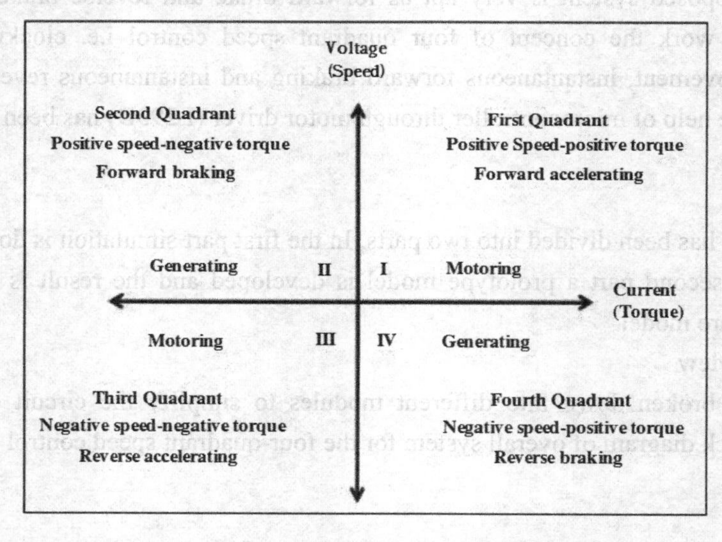

Fig.14.2 Four Quadrants of Operation

3) Pulse Width Modulation

Pulse width Modulation (PWM) is the term used to describe using a digital signal to generate an analogue output signal. PWM is one of the powerful techniques used in control systems today. This is usually used to control the average power to a load in a motor speed control circuit. It is used in wide range of application which includes: speed control, power control, measurement and communication.

Pulse-width modulation (PWM) is a commonly used technique for controlling power to an electrical device, made practical by modern electronic power switches. The average value of voltage (and current) fed to the load is controlled by turning the switch between supply and load on and off at a fast pace. The term duty cycle describes the proportion of on time to the regular interval or period of time, a low duty cycle corresponds to low power, because the power is off for most of the time. Duty cycle is expressed in percent, 100% being fully on.

The main advantage of PWM is that power loss in the switching devices is very low. When a switch is off there is practically no current, and when it is on, there is almost no voltage drop across the switch. Power loss, being the product of voltage and current, is thus in both cases close to zero[3]. PWM works also well with digital controls, which, because of their on/off nature, can easily set the needed duty cycle. The duty cycle determines the speed of the motor. The desired speed can be obtained by changing the duty cycle. The PWM in microcontroller is used to control the duty cycle of DC motor. The PWM pulses generated from the microcontroller are viewed for various duty cycles in the simulation done in proteous software.

4) Motor Driver IC

L293D is a dual H-bridge motor driver integrated circuit (IC). Motor drivers act as current amplifiers since they take a low-current control signal and provide a higher-current signal. This higher current signal is used to drive the motors. L293D contains two inbuilt H-bridge driver circuits. In its common mode of operation, two DC motors can be driven simultaneously, both in forward and reverse direction. The motor operations of two motors can be controlled by input logic at pins 2 & 7 and 10 & 15. Input logic 00 or 11 will stop the corresponding motor. Logic 01 and 10 will rotate it in clockwise and anticlockwise directions, respectively. Enable pins 1 and 9 (corresponding to the two motors) must be high for motors to start operating. When an enable input is high, the associated driver gets enabled. As a result, the outputs become active and work in phase with their inputs. Similarly, when the enable input is low, that driver is disabled, and their outputs are off and in the high-impedance state.

Complete Drive System

To implement this project work three software have used. These are:

(1) Keil.

(2) Proteus.

(3) Flash Magic.

Keil:

Keil compiler has been used to convert high level language into Hex code.

Proteus:

It has been used to simulate the result in software.

Positron Boot Loader:

It has been used to burn Hex code into microcontroller.

The overall block of the system is implemented in the proteus software and the response and the operation of the motor is viewed as in Fig.14.3.

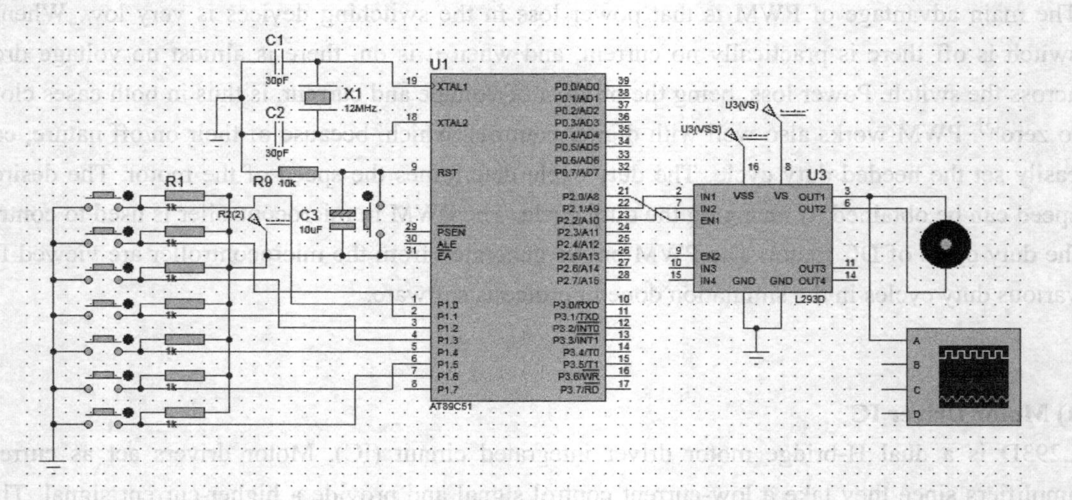

Fig.14.3 System Tested In Software

The response of the motor connected can be seen visually according to the program fed into the microcontroller and the operations are carried accordingly. It is the easiest way to check whether the hardware will get the desired output. The changes can be made to get the desired output and the operation can be carried out accordingly.

Hardware Description

The following procedures are carried out for the for the four quadrant DC motor speed control operation using microcontroller. Here seven switches are interfaced to MC to control the speed of motor in four quadrants. When start switch is pressed the motor starts rotating in full speed being driven by a motor driver IC L293D that receives control signal continuously from the microcontroller. When clockwise switch is pressed the motor rotates in forward direction as per the logic provided by the program from the microcontroller to the motor driver IC. While forward brake is pressed a reverse voltage is applied to the motor by the motor driver IC by sensing reverse logic sent by the microcontroller for a short time period due to and reverse brake switch is pressed the microcontroller delivers a logic to the motor driver IC that develops for very small time a reverse voltage across the running motor due to which instantaneous brake situation happens to the motor. PWM switch is used to rotate the motor at varying speed by delivering from the microcontroller a varying duty cycle to the enable pin of the motor driver IC. It starts from 100% duty cycle and reduces in steps of 10% when it is pressed again and finally reaches to 10% duty cycle and the process repeats. Stop button is used to switch OFF the motor by driving the enable pin to ground from the microcontroller command accordingly.

Practical Implementation

The practical implementation of the four quadrant control of the DC motor is shown in Fig.14.4.

Fig.14.4 Complete Prototype Hardware Model

Words

microcontroller	*n.* 微控制器，单片机
MCU	Microcontroller Unit (微控制器，单片机)
DC	Direct Current (直流电)
quadrant	*n.* 象限
characteristics.	*n.* 特性；特征
brake	*n.* 制动器
driver	*n.* 驱动器，驱动程序
methodology	*n.* 方法学，方法论
AC	Alternating Current (交流电)

power supply	电源
transformer	*n.* 变压器
rectifier	*n.* 整流，整流器
voltage regulator	调压器，稳压器
IC	Integrated Circuit（集成电路，芯片）
emf	Electromotive Force（电动势）
PWM	Pulse Width Modulation（脉冲宽度调制）
load	*n.* 负载；*v.* 加载
duty cycle	占空比
current	*n.*（水、气、电）流
power	*n.* 功率
product	*n.* 乘积
in phase with	与……同相位
impedance	*n.* 阻抗
spatial visualization	空间想象
horizontal projection	水平投影
frontal projection	正投影
profile projection	侧投影
quadrant	*n.* 象限
prototype	*n.* 原型，样机
composite object	组合体

Notes

[1] In this work the concept of four quadrant speed control i.e. clockwise movement, anticlockwise movement, instantaneous forward braking and instantaneous reverse braking of a dc motor with the help of microcontroller through motor driver (L293D) has been proposed.

【译文】本文提出了利用单片机通过电机驱动芯片(L293D)，实现直流电动机的顺时针、逆时针、瞬时正向制动和瞬时反向制动的四象限速度控制思想。

[2] The circuit uses standard power supply comprising of a step-down transformer from 230V to 12V and the four diodes forming a bridge rectifier that delivers pulsating dc which is unregulated is regulated to constant 5V DC.

【译文】该电路采用标准电源，由 230V 到 12V 的降压变压器和由四个二极管组成的桥式整流器构成，桥式整流器输出的脉动直流电稳压成恒定的 5V 直流电。

[3] When a switch is off there is practically no current, and when it is on, there is almost no voltage drop across the switch. Power loss, being the product of voltage and current, is thus in both cases

close to zero.

【译文】当开关关闭时，几乎没有电流通过，当开关打开时，开关上几乎没有电压降。因此，作为电压和电流的乘积，功率损耗在这两种情况下都接近于零。

Part B Reading Materials

With today's advanced VLSI technologies, microcontrollers are now far cheaper, faster, lower power, and more powerful than their underpowered predecessors. Designers now have more choices than yesterday's simple 8-bit microprocessors when small, low power systems are warranted.

The period between 1970 and 1980 saw an explosion in microprocessor technology. The room-sized computers of the 1950's and 1960's were fast being replaced with vastly smaller and more reliable LSI chips. These 4- and 8-bit microprocessors were a giant step for the computer industry, and their impact is still being felt today.

Following these 4 and 8-bit microprocessors were the vastly more expensive 16 and 32-bit architectures. Usually reserved for supercomputer and minicomputer applications, they were high power, high cost devices which required hundreds of support chips to operate.

At the time, all such microprocessors required support chips such as SRAM, ROM, I/O, and glue logic. The concept of "single-chip" operation did not exist. These chips had the address/data/control bus structure, and resultingly could not operate on their own.

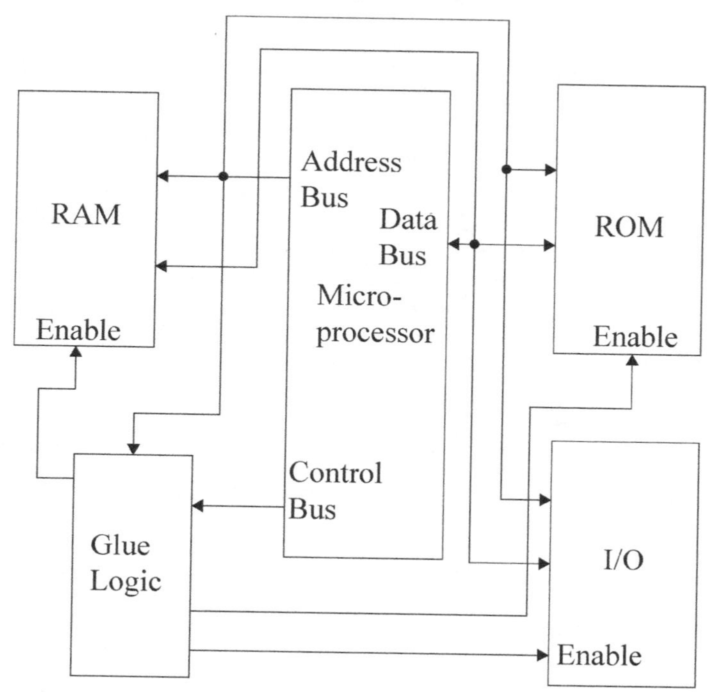

Fig.14.5 Block Diagram of a Typical Microprocessor-based System

Modern microprocessors still retain this RAM/ROM/IO structure along with data and address busses. This configuration allows the most expandability and allows the highest performance when designing large systems. However, in the realm of embedded systems, a new trend in processor design has developed, called the microcontroller.

Microcontrollers were developed out of the need for small, low power systems. Microcontrollers typically do not have the expandability or performance that microprocessors have. They are designed with control and consumer applications in mind, such as data logging, appliances, personal electronic devices such as walkmans and digital watches, etc. In the past, when a designer needed to design the electrical interface for a microwave, it was done with dedicated hardware. These days such control electronics are completely replaced with a small, fast, and cheap microcontroller. This allows software upgradability and modularity of design. When the company decides to design their next microwave, they can use all the same hardware only needing to change the software. The typical BMW automobile can contain 70+ 68HC11 microcontrollers.

Part C　Exercises

A brief introduction to the popular microcontroller 8051.

Lesson 15 Mechanical-electrical Integration System Design

教学目的和要求

通过本文的学习，了解机电系统集成设计的含义、方法、步骤；要求掌握设计中常用的机械部件及电子元件的名称及其属性的专业表达。

重点和难点

（1）了解机电系统集成设计的含义。
（2）重点掌握机电系统设计中常用机械部件及电子元件的名称。

Part A Text

Introduction

Cross-disciplinary integration of mechanical engineering, electrical and electronic engineering as well as recent advances in information engineering are becoming more and more crucial for future collaborative design, manufacture, and maintenance of a wide range of engineering products and processes[1]. In order to allow for additional synergy effects in collaborative product creation, designers from all disciplines involved need to adopt new approaches to design, which facilitate concurrent cross-disciplinary collaboration in an integrated fashion.

Mechatronic systems usually encompass mechanical, electronic, electrical, and software components (Fig.15.1). The design of mechanical components requires a sound understanding of core mechanical engineering subjects, including mechanical devices and engineering mechanics. For example, expertise regarding lubricants, heat transfer, vibrations, and fluid mechanics are only a few aspects to be considered for the design of most mechatronic systems. Mechanical devices include simple latches, locks, ratchets, gear drives and wedge devices as well as complex devices such as harmonic drives and crank mechanisms. Engineering mechanics is concerned with the kinematics and dynamics of machine elements. Kinematics determines the position, velocity, and acceleration of machine links. Kinematic analysis is used to find the impact and jerk on a machine element. Dynamic analysis is used to determine torque and force required for the motion of link in a mechanism. In dynamic analysis, friction and inertia play an important role.

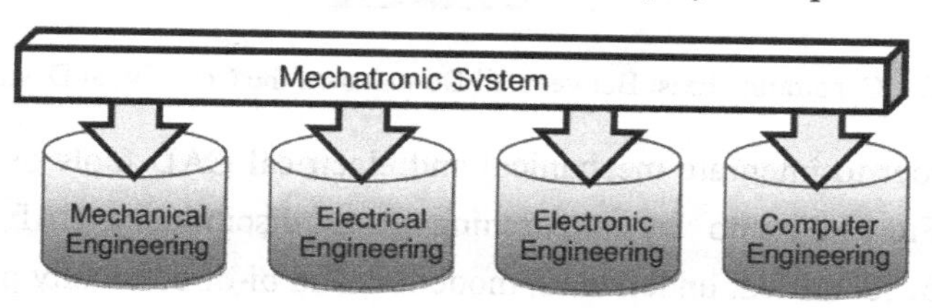

Fig.15.1 The scope of mechatronic system

Electronics involves measurement systems, actuators, and power control. Measurement systems in general comprise of three elements: sensors, signal conditioners, and display units. A sensor responds to the quantity being measured from the given electrical signal, a signal conditioner takes the signal from the sensor and manipulates it into conditions suitable for display, and in the display unit the output from the signal conditioner is displayed. Actuation systems comprise the elements which are responsible for transforming the output from the control system into the controlling action of a machine or a device. And finally power electronic devices are important in the control of power-operated devices. The silicon controlled rectifier is an example of a power electronic device which is used to control dc motor drives.

The ideal process of concurrent or simultaneous engineering is characterized by parallel work of a potentially distributed community of designers that know about the parallel work of their colleagues and collaborate as necessary. In order to embed the discussion aspect, mechatronic products should be designed in an integrated fashion that allows for designers of both electrical and mechanical engineering domains to automatically receive feedback regarding design modifications made on either side throughout the design process. This means, that if a design modification of a mechanical component of a mechatronic system will lead to a design modification of an electrical aspect of the mechatronic system or vice versa, the engineer working at the counterpart system should get notified as soon as possible[2]. Obviously, even on conceptual design level, mechanical and electrical design aspects of mechatronic systems are highly intertwined through a substantial number of constraints existing between their components (Fig.15.2).

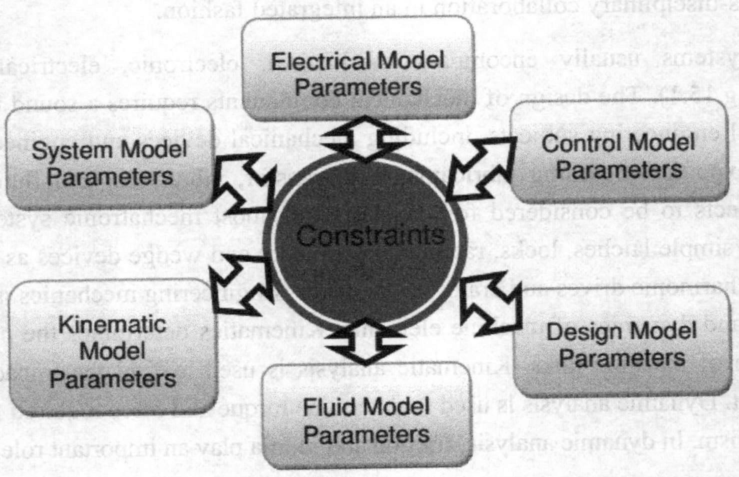

Fig.15.2 Constraints Exist Between all Domains on the Conceptual Design Level

Consequently, in order to integrate mechanical and electrical CAD tools on systems realization/ integration level (Fig.15.3) into an overarching cross disciplinary CAE environment, these constraints have to be identified, understood, modelled, and bi-directionally processed[3].

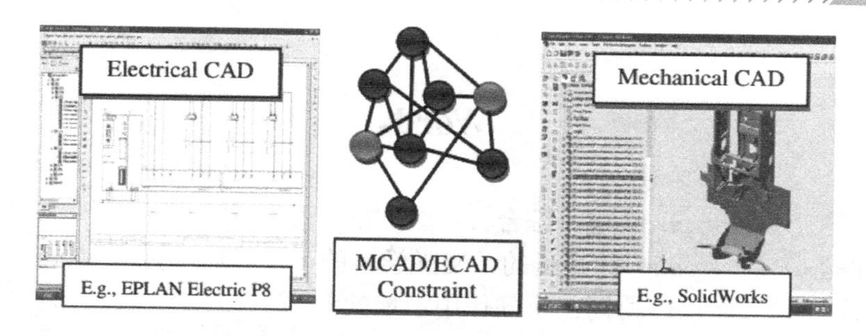

Fig.15.3 Constraints Exist between MCAD and ECAD Models on System Realization Level

Modelling Mechatronic Systems

The constraint modelling-based approach proposed in this chapter (Fig.15.4) is similar to the semantic network approach in that constraints are being modelled as nodes and relationships are drawn between nodes. The components of mechatronic systems are modelled as objects with attributes, and constraints between these attributes are identified and modelled. The procedures of the proposed constraint modeling approach are as follows:

STEP 1: List all components of the mechatronic system and their attributes and classify the components in either the mechanical domain or the electrical domain.

STEP 2: Based on the attributes of the component, draw the constraint relationship between the components in the domain and appropriately label the constraint by the constraint categories.

STEP 3: Based on the attributes of the component, draw the constraint relationship between the components across the domains and appropriately label the constraint by the constraint categories.

STEP 4: Construct a table of constraints for the particular mechatronic system. The table contains a complete list of the every component of the mechatronic system, the table is to indicate that, when a particular attribute of the component is being modified, which attribute of which component (both within the domain and across the domain) would be affected.

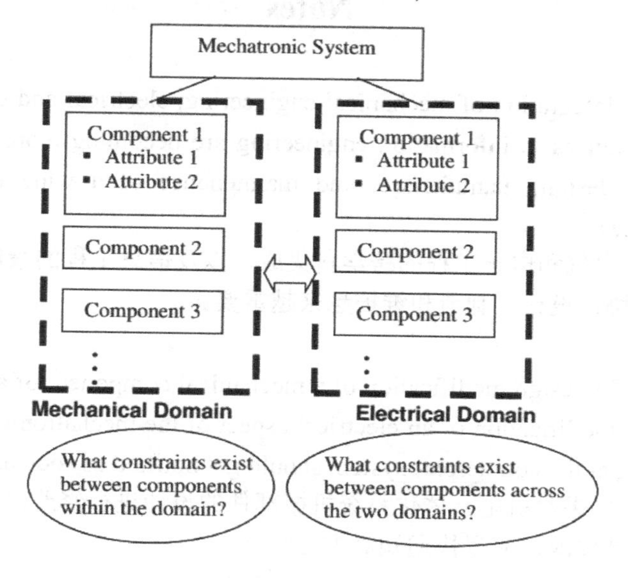

Fig.15.4 A Graphical View of the Constraint Modelling Approach

Words

integration	*n*. 集成；综合
cross-disciplinary	*adj*. 跨学科的；交叉学科的
synergy	*n*. 协同；协同作用；增效
engineering mechanics	工程力学
lubricant	*n*. 润滑油，润滑剂
vibration	*n*. 振动
fluid mechanics	流体力学
harmonic	*n*. 谐波
kinematics	*n*. 运动学
dynamics	*n*. 动力学
velocity	*n*. 速度
actuator	*n*. 执行机构
sensor	*n*. 传感器
signal conditioner	信号调节器
Silicon controlled	可控硅
conceptual design	概念设计，方案设计
intertwined	*adj*. 缠绕的；错综复杂的
attribute	*n*. 属性；特质
constraint	*n*. 约束

Notes

[1] Cross-disciplinary integration of mechanical engineering, electrical and electronic engineering as well as recent advances in information engineering are becoming more and more crucial for future collaborative design, manufacture, and maintenance of a wide range of engineering products and processes.

【译文】机械工程、电气和电子工程的跨学科集成，以及信息工程的最新进展，对未来各种工程产品和过程的协作设计、制造和维护越来越重要。

[2] This means, that if a design modification of a mechanical component of a mechatronic systems will lead to a design modification of an electrical aspect of the mechatronic system or vice versa, the engineer working at the counterpart system should get notified as soon as possible.

【译文】这意味着，如果对机电一体化系统机械部件的设计进行修改，就将导致机电一体化系统电气方面的设计修改，应尽快通知对方工程师。

[3] Consequently, in order to integrate mechanical and electrical CAD tools on systems realization/integration level into an overarching cross disciplinary CAE environment, these constraints have to be identified, understood, modelled, and bi-directionally processed.
【译文】因此，为了实现系统集成，将机械和电气 CAD 工具集成到跨学科的 CAE 环境中，必须对这些约束进行识别、理解、建模和双向处理。

Part B Reading Materials

A robot is a mechatronic system capable to replace or assist the human operator in carrying out a variety of physical tasks. The interaction with the surrounding environment is achieved through sensors and transducers and the computer-controlled interaction systems emulate human capabilities. The example investigated is the SG5-UT robot arm designed by Alex Dirks of the CrustCrawler team (Fig.15. 5).

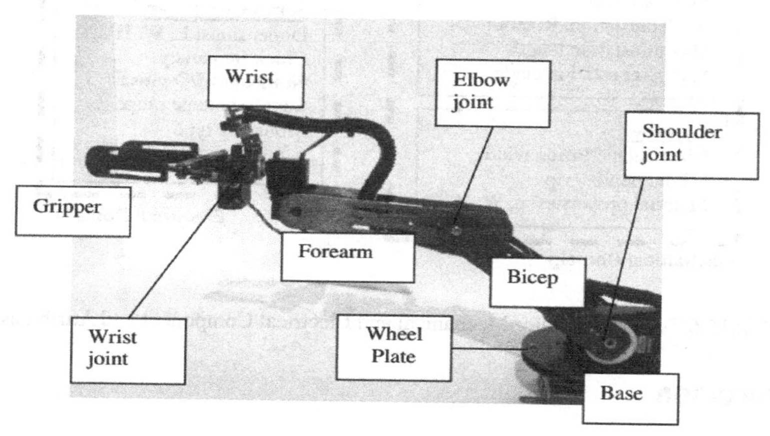

Fig.15.5 The CrustCrawler SG%-UT Robot Arm

List of major mechanical component:

- Base and Wheel plates
- Links: Bicep, Forearm, Wrist
- Gripper o Joints: shoulder, elbow, and wrist
- Hitec HS-475HB servos (base, wrist and gripper)
- Hitec HS-645MG servos (elbow bend)
- Hitec HS-805BB servos (shoulder bend)

All components of the SG5-UT robot arm and their attributes and classify the components in either the mechanical domain or the electrical domain. are listed in Fig.15.6.

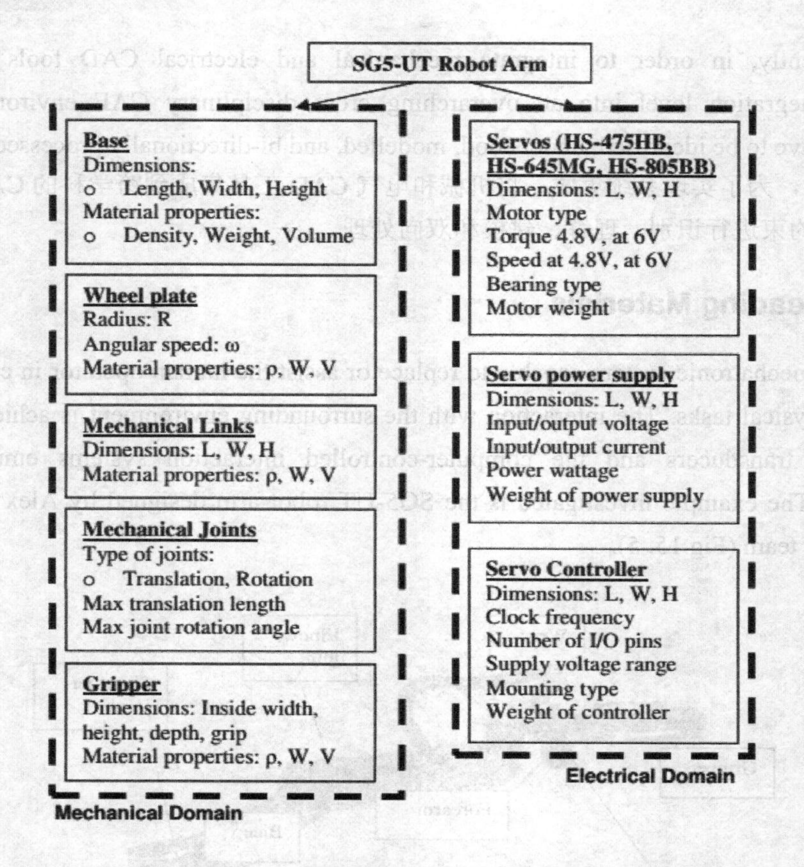

Fig.15.6 A List of Major Mechanical and Electrical Component and Attributes

Part C Exercises

List the main constraints considered in mechanical-electrical integration system design with examples.

Lesson 16 Principle and Maintenance of Numerical Control

教学目的和要求

通过本文的学习，了解数控技术的概念、原理和应用；要求掌握常见数控的维修，了解数控技术在实际生活中的应用。

重点和难点

（1）了解数控技术在机械制造中的应用。

（2）掌握数控技术的专业词汇及其英文表达。

Part A Text

Computer numerical control (CNC) is the automation of machine tools by means of computers executing pre-programmed sequences of machine control commands. This is in contrast to machines that are manually controlled by hand wheels or levers, or mechanically automated by cams alone.

In modern CNC systems, the design of a mechanical part and its manufacturing program is highly automated. The part's mechanical dimensions are defined using computer-aided design (CAD) software[1], and then translated into manufacturing directives by computer-aided manufacturing (CAM) software. The resulting directives are transformed (by "post processor" software) into the specific commands necessary for a particular machine to produce the component, and then are loaded into the CNC machine.

Since any particular component might require the use of a number of different tools—drills, saws, etc.—modern machines often combine multiple tools into a single "cell". In other installations, a number of different machines are used with an external controller and human or robotic operators that move the component from machine to machine. In either case, the series of steps needed to produce any part is highly automated and produces a part that closely matches the original CAD.

Motion is controlled along multiple axes, normally at least two (X and Y), and a tool spindle that moves in the Z (depth)[2]. The position of the tool is driven by direct-drive stepper motor or servo motors in order to provide highly accurate movements, or in older designs, motors through a series of step-down gears. Open-loop control works as long as the forces are kept small enough and speeds are not too great. On commercial metalworking machines, closed loop controls are standard and required in order to provide the accuracy, speed, and repeatability demanded.

As the controller hardware evolved, the mills themselves also evolved. One change has been to enclose the entire mechanism in a large box as a safety measure, often with additional safety interlocks to ensure the operator is far enough from the working piece for safe operation. Most new CNC systems built today are 100% electronically controlled.

CNC-like systems are now used for any process that can be described as a series of movements and operations. These include laser cutting, welding, friction stir welding, ultrasonic welding, flame and plasma cutting, bending, spinning, hole-punching, pinning, gluing, fabric cutting, sewing, tape and fiber placement, routing, picking and placing, and sawing[3]. In numerical control systems, the position of the tool is defined by a set of instructions called the part program.

Numerical precision and equipment backlash

Within the numerical systems of CNC programming it is possible for the code generator to assume that the controlled mechanism is always perfectly accurate, or that precision tolerances are identical for all cutting or movement directions. This is not always a true condition of CNC tools. CNC tools with a large amount of mechanical backlash can still be highly precise if the drive or cutting mechanism is only driven so as to apply cutting force from one direction, and all driving systems are pressed tightly together in that one cutting direction[4]. However a CNC device with high backlash and a dull cutting tool can lead to cutter chatter and possible workpiece gouging. Backlash also affects precision of some operations involving axis movement reversals during cutting, such as the milling of a circle, where axis motion is sinusoidal. However, this can be compensated for if the amount of backlash is precisely known by linear encoders or manual measurement.

The high backlash mechanism itself is not necessarily relied on to be repeatedly precise for the cutting process, but some other reference object or precision surface may be used to zero the mechanism, by tightly applying pressure against the reference and setting that as the zero reference for all following CNC-encoded motions[5]. This is similar to the manual machine tool method of clamping a micrometer onto a reference beam and adjusting the Vernier dial to zero using that object as the reference.

Positioning control system

Positioning control is handled by means of either an open loop or a closed loop system. In an open loop system, communication takes place in one direction only: from the controller to the motor. In a closed loop system, feedback is provided to the controller so that it can correct for errors in position, velocity, and acceleration, which can arise due to variations in load or temperature. Open loop systems are generally cheaper but less accurate. Stepper motors can be used in both types of systems, while servo motors can only be used in closed systems.

Coding

Having the correct Speeds and Feeds in the program provides for a more efficient and smoother product run. Incorrect speeds and feeds will cause damage to the tool, machine spindle and even the product. The quickest and simplest way to find these numbers would be to use a calculator that can be found online. A formula can also be used to calculate the proper speeds and feeds for a material. This values can be found online or Machinery's Handbook.

Tool / machine crashing

In CNC, a "crash" occurs when the machine moves in such a way that is harmful to the machine, tools, or parts being machined, sometimes resulting in bending or breakage of cutting tools, accessory clamps, vises, and fixtures, or causing damage to the machine itself by bending guide rails, breaking drive screws, or causing structural components to crack or deform under strain[6]. A mild crash may not damage the machine or tools, but may damage the part being machined so that it must be scrapped.

Many CNC tools have no inherent sense of the absolute position of the table or tools when turned on. They must be manually "homed" or "zeroed" to have any reference to work from, and these limits are just for figuring out the location of the part to work with it, and aren't really any sort of hard motion limit on the mechanism. It is often possible to drive the machine outside the physical bounds of its drive mechanism, resulting in a collision with itself or damage to the drive mechanism. Many machines implement control parameters limiting axis motion past a certain limit in addition to physical limit switches. However, these parameters can often be changed by the operator.

Many CNC tools also don't know anything about their working environment. Machines may have load sensing systems on spindle and axis drives, but some do not. They blindly follow the machining code provided and it is up to an operator to detect if a crash is either occurring or about to occur, and for the operator to manually abort the active process. Machines equipped with load sensors can stop axis or spindle movement in response to an overload condition, but this does not prevent a crash from occurring. It may only limit the damage resulting from the crash. Some crashes may not ever overload any axis or spindle drives.

If the drive system is weaker than the machine structural integrity, then the drive system simply pushes against the obstruction and the drive motors "slip in place". The machine tool may not detect the collision or the slipping, so for example the tool should now be at 210 mm on the X axis, but is, in fact, at 32mm where it hit the obstruction and kept slipping[7]. All of the next tool motions will be off by 178mm on the X axis, and all future motions are now invalid, which may result in further collisions with clamps, vises, or the machine itself. This is common in open loop stepper systems, but is not possible in closed loop systems unless mechanical slippage between the motor and drive mechanism has occurred. Instead, in a closed loop system, the machine will continue to attempt to move against the load until either the drive motor goes into an overcurrent condition or a servo following error alarm is generated.

Collision detection and avoidance is possible, through the use of absolute position sensors (optical encoder strips or disks) to verify that motion occurred, or torque sensors or power-draw sensors on the drive system to detect abnormal strain when the machine should just be moving and not cutting, but these are not a common component of most hobby CNC tools[8].

Instead, most hobby CNC tools simply rely on the assumed accuracy of stepper motors that rotate a specific number of degrees in response to magnetic field changes. It is often assumed the stepper is perfectly accurate and never missteps, so tool position monitoring simply involves counting the number of pulses sent to the stepper over time. An alternate means of stepper position monitoring is usually not available, so crash or slip detection is not possible.

Commercial CNC metalworking machines use closed loop feedback controls for axis movement. In a closed loop system, the control is aware of the actual position of the axis at all times. With proper control programming, this will reduce the possibility of a crash, but it is still up to the operator and programmer to ensure that the machine is operated in a safe manner. However, during the 2000s and 2010s, the software for machining simulation has been maturing rapidly, and it is no longer uncommon for the entire machine tool envelope (including all axes, spindles, chucks, turrets, tool holders, tailstocks, fixtures, clamps, and stock) to be modeled accurately with 3D solid models, which allows the simulation software to predict fairly accurately whether a cycle will involve a crash[9]. Although such simulation is not new, its accuracy and market penetration are changing considerably because of computing advancements.

Words

pre-programmed	预先设定
dimensions	*n*. 尺寸
aided design	辅助设计
post processo	后处理
cell	*n*. 单元
servo motors	*n*. 伺服电机
open-loop	*adj*. 开环的
compensated	*v*. 补偿

Notes

[1] The part's mechanical dimensions are defined using computer-aided design (CAD) software, and then translated into manufacturing directives by computer-aided manufacturing (CAM) software.
【译文】零件的尺寸采用计算机辅助设计（CAD）软件定义，然后通过计算机辅助制造（CAM）软件转化为制造指令。

[2] Motion is controlled along multiple axes, normally at least two (X and Y), and a tool spindle that moves in the Z (depth).
【译文】多轴控制运动通常至少有两个轴(X 轴和 Y 轴)参与，还有在 Z(深度)方向运动的工具主轴。

[3] These include laser cutting, welding, friction stir welding, ultrasonic welding, flame and plasma cutting, bending, spinning, hole-punching, pinning, gluing, fabric cutting, sewing, tape and fiber placement, routing, picking and placing, and sawing.
【译文】包括激光切割、焊接、搅拌摩擦焊接、超声波焊接、火焰，以及等离子切割、弯曲、纺丝、打孔、钉扎、胶合、织物切割、缝纫、胶带和纤维放置、布线、采摘和放置及锯切。

[4] CNC tools with a large amount of mechanical backlash can still be highly precise if the drive or cutting mechanism is only driven so as to apply cutting force from one direction, and all driving systems are pressed tightly together in that one cutting direction.
【译文】驱动机构或切削机构如果只从一个方向施加切削力，那么数控刀具的机械间隙虽然较大，但是仍可以达到很高的精度，并且所有的驱动系统都被紧紧地压在同一切削方向上。

[5] The high backlash mechanism itself is not necessarily relied on to be repeatedly precise for the cutting process, but some other reference object or precision surface may be used to zero the mechanism, by tightly applying pressure against the reference and setting that as the zero reference for all following CNC-encoded motions.
【译文】高间隙机构本身不一定要在切削过程中重复精确定位，但可以使用其他参考物体或精密表面来实现机构的零点，方法是对基准施加压力，并将其设置为所有后续数控编码运动的零参考。

[6] In CNC, a "crash" occurs when the machine moves in such a way that is harmful to the machine, tools, or parts being machined, sometimes resulting in bending or breakage of cutting tools, accessory clamps, vises, and fixtures, or causing damage to the machine itself by bending guide rails, breaking drive screws, or causing structural components to crack or deform under strain.

【译文】在数控系统中，当机器以对被加工的机床、工具或零件有害的方式移动时，就会发生“崩溃”，有时导致切削工具、附件夹子、黏滞和夹具弯曲或断裂，或者通过弯曲导轨时折断驱动螺钉或导致结构部件在应变下开裂或变形而对机器造成损害。

[7] The machine tool may not detect the collision or the slipping, so for example the tool should now be at 210 mm on the X axis, but is, in fact, at 32mm where it hit the obstruction and kept slipping.

【译文】机床可能检测不到碰撞或滑动，例如，刀具现在位于 X 轴的 210mm 处，但实际上在 32mm 处就碰到了障碍物，然后仍继续滑动。

[8] Collision detection and avoidance is possible, through the use of absolute position sensors (optical encoder strips or disks) to verify that motion occurred, or torque sensors or power-draw sensors on the drive system to detect abnormal strain when the machine should just be moving and not cutting, but these are not a common component of most hobby CNC tools.

【译文】避免和检测碰撞是可能的，通过使用绝对位置传感器（光学编码器条或磁盘）来验证是否发生了运动，或者使用驱动系统上的扭矩传感器或功率吸引传感器来检测机器是否只是移动而不是切割时的异常应变，但这些并不是大多数数控业余工具的常见组成部分。

[9] However, during the 2000s and 2010s, the software for machining simulation has been maturing rapidly, and it is no longer uncommon for the entire machine tool envelope (including all axes, spindles, chucks, turrets, tool holders, tailstocks, fixtures, clamps, and stock) to be modeled accurately with 3D solid models, which allows the simulation software to predict fairly accurately whether a cycle will involve a crash.

【译文】然而，在 2000 年代和 2010 年代，机械加工仿真软件已经迅速成熟，整个机床（包括所有轴、主轴、卡盘、转塔、刀具架、尾柱、夹具和库存）都能用三维实体模型精确建模，这使仿真软件能够相当准确地预测一个运动周期是否会发生碰撞。

Part B　Reading Materials

CNC system and related automation products mainly for supporting CNC machine tools. NC machine tool is a mechanical and electrical integration product formed by the infiltration of the new technology represented by the numerical control system to the traditional mechanical manufacturing industry. The machine tool equipped with the numerical control system greatly improves the precision, speed and efficiency of the parts processing. This NC machine tool is the national industrial modernization of the important material basis.

Part C　Exercises

Give a brief account of your understanding of CNC.

Lesson 17 C Language Programming Example

教学目的和要求

通过本文的学习，了解单片机C语言程序的基本结构、组成，掌握单片机C语言程序设计涉及的主要变量类型、运算符的分类及其表达方式。

重点和难点

（1）了解单片机C语言程序的基本组成。
（2）重点掌握单片机C语言程序设计运算符的含义及分类。

Part A Text

Since the heart of an embedded control system is a microcontroller, we need to be able to develop a program of instructions for the microcontroller to use while it controls the system in which it is embedded.

For this embedded microcontroller, we will be using a programming language called “C”. C is extremely flexible, and allows programmers to perform many low-level functions which are not easily accessible in languages like FORTRAN or Pascal[1]. Unfortunately, the flexibility of C also makes it easier for the programmer to make mistakes and potentially introduce errors into their program. To avoid this, you should be very careful to organize your program so that it is easy to follow, with many comments so that you and others can find mistakes quickly.

A Simple Program in C

The following program is similar to the first programming example used in most C programming books and illustrates the most basic elements of a C program.

```
#include <stdio.h> /* include file */
main() /* begin program here */
{ /* begin program block */
printf("Hello World\n");
/* send Hello World
to the terminal */
} /* end the program block */
```

The first line instructs the compiler to include the *header file* for the standard input/output functions. This line indicates that some of the functions used in the file (such as printf) are not defined in the file, but instead are in an external library (stdio.h is a standard library header file).

This line also illustrates the use of comments in C. Comments begin with the two character sequence "/*" and end with the sequence "*/". Everything between is ignored and treated as comments by the compiler. Nested comments are not allowed.

The second line in the program is `main()`. Every C program contains a function called `main()` which is the function that executes first. The next line is a curly bracket. Paired curly brackets are used in C to indicate a *block* of code. In the case above, the block belongs to the `main()` statement preceding it. The `printf` line is the only statement inside the program. In C, programs are broken up into *functions*.

The `printf` function sends text to the terminal. In our case, the C8051 will send this text over the serial port to a *computer terminal*, where we can view it. (You will use software on your laptop to simulate a terminal.) This line also illustrates that every statement in C is followed by a semicolon. The compiler interprets the semicolon as the end of one statement, and then allows a new statement to begin.

You may also notice that the comment after the `printf` statement continues over more than one line. It is important to remember that *everything* between the comment markers is ignored by the compiler.
The last line is the closing curly bracket which ends the block belonging to the main function.

Syntax Specifics
There are many syntax rules in C. We explain the basics, along with the specifics of the *SDCC C Compiler* which are not in your textbook. Additional information about the C language can be found in the tutorials, and in any C reference text.

Declarations
One thing which was distinctly missing from the first example program was a variable[2]. The type of variables available with the *SDCC C Compiler* for the C8051 microcontroller and their declaration types are listed below in Table 17.1.

Table 17.1 *SDCC C Compiler* variable types

Type[a]	Size (bytes)	Smallest Value	Largest Value
integer			
(unsigned) char	1	0	255
signed char	1	−128	127
(signed) short	2	−32768	32767
unsigned short	2	0	65535
(signed) int	2	−32768	32767
unsigned int	2	0	65535

Continue

Type[a]	Size (bytes)	Smallest Value	Largest Value
(signed) long	4	-2147483648	2147483647
unsigned long	4	0	4294967295
floating point			
float	4	1.2 × 1038	1.2 × 1038
SDCC specific			
bit	1/8 = 1 bit	0	1
sbit	1/8 = 1 bit	0	1

a. The items in parentheses are not required, but are implied by the definition. We recommend that you state these definitions explicitly to avoid errors due to misdefinition.

The format for declaring a variable in a C program is as follows:

<type> `variablename;`

For example, the line

```
int i;
```

would declare an integer variable `i`. Although there are a large variety of variable types available, it is important to realize that the larger the size of the data type, the more time will be required by the C8051 to make the calculations. Increased calculation time is also an important consideration when using floating-point variables. It is suggested that in the interest of keeping programs small and efficient, you should not use floating point numbers unless absolutely necessary.

Repetitive Structures

Computers are very useful when repeating a specific task, and almost every program utilizes this capability. The repetitive structures `for`, `while`, and `do..while` are all offered by C.

for Loops

The most common of looping structures is the `for` loop, which looks like this

```
for(initialize_statement; condition; increment){
...
}
```

In the example above, the `for` loop will perform the *`initialize_statement`* one time before commencing the loop. The *`condition`* will then be checked to make sure that it is true (non-zero). As long as the *`condition`* is true the statements within the loop block will be performed. After each iteration of the loop, the increment statement is performed. For example:

```
for(i=0; i<10; i++) {
display(i);
}
```

The statement above will initially set i equal to zero, and then call a user-defined function named display()10 times. Each time through the loop, the value of i is incremented by one. After the tenth time through, i is set to 10, and the for loop is ended since i is not less than ten.

While Loops

Another frequently used loop structure is the while loop, which follows this format

```
while(condition){
...
}
```

When a while loop is encountered, the condition given in parenthesis is evaluated. If the condition is true (evaluates to non-zero), the statements inside the braces are executed. After the last of these statements is executed, the condition is evaluated again, and the process is repeated.

When the condition is false (evaluates to zero), the statements inside the braces are skipped over, and execution continues after the closing brace. As an example, consider the following:

```
i = 0;
while(i<10) {
i++;
display(i);
}
```

The above while loop will give the same results as the preceding example given with the for loop. The variable i is first initialized to zero. When the while is encountered in the next line, the computer checks to see if i is less than 10. Since i begins the loop with the value 0 (which is less than ten), the statements inside the braces will be executed. The first line in the loop, i++, increments i by 1 and is equivalent to i=i+1. The second line calls a function named display with the current value of i passed as a parameter. After display is called, the computer returns to the while statement and checks the condition (i < 10). Since after the first iteration of the loop the value of i is 1, the condition (i < 10) evaluates to logical *TRUE* or equivalently *1*, the loop will again be executed. The looping will continue until i equals 10 when the condition (i < 10) will evaluate as being false. The program will then skip over all the statements within the braces of the while construct, and proceed to execute the next statement following the closing brace "}".

Operators

In addition to a full complement of keywords and functions, C also includes a full range of operators. Operators usually have two arguments, and the symbol between them performs an operation on the two arguments, replacing them with the new value. You are probably most familiar with the mathematical operators such as + and -, but you may not be familiar with the bitwise and logical operators which are used in C. Table 17.2 - Table 17.3 list some of the different types of operators available. The operators are also listed in the order of precedence in

Table 17.4. Just as in algebra where multiplication precedes addition, all C operators obey a precedence which is summarized in table Table 17.5.

Mathematical

The symbols used for many of the C mathematical operators are identical to the symbols for standard mathematical operators, e.g., add +, subtract -, and divide / [3]. Table 17.2 lists the mathematical operators.

Table 17.2 Mathematical operators

operator	description
*	multiplication
/	division
%	mod (remainder)
+	addition
-	subtraction

Relational, Equality, and Logical

The C language offers a full range of control structures including `if.else`, `while`, `do..while`, and `switch`. Most of these structures should be familiar from previous computing classes, so the concepts are left to a reference text on C. In C, remember that any non-zero value is true, and a value of zero is false. Relational, equality, and logical operators are used for tests in control structures, and are shown in Table 17.3. All operators in this list have two arguments (one on each side of the operator).

Table 17.3 Relational, equality, and logical operators

operator	description	operator	description
<	less than	==	equal to
>	greater than	!=	not equal to
<=	less than or equal to	||	logical OR
>=	greater than or equal to	&&	logical AND

Bitwise

C can perform some low-level operations such as bit-manipulation that are difficult with other programming languages. In fact, some of these bitwise functions are built into the language. Table 17.4 summarizes the bitwise operations available in C.

Table 17.4 Bitwise and shift operators

operator	description	example	result
&	bitwise AND	0x88 & 0x0F	0x08
^	bitwise XOR	0x0F ^ 0xFF	0xF0
|	bitwise OR	0xCC | 0x0F	0xCF
<<	left shift	0x01 << 4	0x10
>>	right shift	0x80 >> 6	0x02

Unary

Some C operators are meant to operate on one argument, usually the variable immediately following the operator. Table 17.5 gives a list of those operators, along with some example for reference purposes.

Assignment

Most mathematical and bitwise operators in the C language can be combined with an equal sign to create an assignment operator. For example, a+=3 is a statement which will add 3 to the current value of a. This is a very useful shorthand notation for a=a+3;. All of the assignment operators have the same precedence as equals, and are listed in the precedence table 17.5.

Table 17.5 Unary operators

operator	description	example	equivalent
++	post-increment	j = i++;	j = i; i = i + 1;
++	pre-increment	j = ++i;	i = i + 1; j = i;
--	post-decrement	j = i--;	j = i; i = i - 1;
--	pre-decrement	j = --i;	i = i - 1; j = i;
*	pointer dereference	*ptr	value at a memory location whose address is in ptr
&	reference (pointer) of	&i	the address of i
+	unary plus	+i	i
-	unary minus	-i	the negative of i
~	ones complement	~0xFF	0x00
!	logical negation	!(0)	(1)

Precedence and Associativity

All operators have a precedence and an associativity. Table 17.6 illustrates the precedence of all operators in the language. Operators on the same row have equal precedence, and precedence decreases as you move down the table.

Table 17.6 Operator precedence and associativity

operators[a]	associativity
() [] ->.	left to right
! ~ ++ -- * & (type) size of	right to left
* / %	left to right
+ -	left to right
<< >>	left to right
< <= > >=	left to right

Continue

<table>
<tr><th>operators[a]</th><th>associativity</th></tr>
<tr><td>== !=</td><td>left to right</td></tr>
<tr><td>&</td><td>left to right</td></tr>
<tr><td>^</td><td>left to right</td></tr>
<tr><td>|</td><td>left to right</td></tr>
<tr><td>&&</td><td>left to right</td></tr>
<tr><td>||</td><td>left to right</td></tr>
<tr><td>?:</td><td>right to left</td></tr>
<tr><td>= += -= *= /= %= &= ^= |= <<= >>=</td><td>right to left</td></tr>
<tr><td>,</td><td>left to right</td></tr>
</table>

a. This table is from Kernighan and Ritchie, *The C Programming Language*.

Words

compiler	*n.* 编译器
header file	头文件
external library	外部类库
comment	*n.* 注释
nested	*n.* 嵌套的
function	*n.* 函数
curly bracket	花括号，波形括号
block of code	程序块
serial port	串行口
semicolon	*n.* 分号
specific	*n.* 特性，细节
SDCC	Small Device C Compiler（C 语言小型编译器）
declaration	*n.* 声明
variable	*n.* 变量
repetitive structure	重复结构
loop	*n.* 循环
iteration	*n.* 迭代
parenthesis	*n.* 圆括号
brace	*n.* 大括号
operator	*n.* 运算符，操作符
precedence	*n.* 优先权，优先级
mathematical operator	算术运算符
logical operator	逻辑运算符
bitwise	*n.* 按位
shift	*n.* 移位

unary operator	一元运算符，单目运算符
assignment	*n.* 赋值

Notes

[1] C is extremely flexible, and allows programmers to perform many low-level functions which are not easily accessible in languages like FORTRAN or Pascal.
【译文】C 语言非常灵活，允许程序员执行许多低级函数，这些函数在 FORTRAN 或 Pascal 语言中不容易访问。

[2] One thing which was distinctly missing from the first example program was a variable.
【译文】第一个示例程序中明显遗漏了一个变量。

[3] The symbols used for many of the C mathematical operators are identical to the symbols for standard mathematical operators, e.g., add +, subtract -, and divide /.
【译文】许多 C 算术运算符所使用的符号与标准算术运算符的符号相同，例如，加号“+”、减号“-”、除号“/”。

Part B Reading Materials

Elements of a Program

If you are going to construct a building, you need two things: the bricks and a blueprint that tells you how to put them together. In computer programming, you need two things: data (variables) and instructions (code or functions). Variables are the basic building blocks of a program. Instructions tell the computer what to do with the variables. Comments are used to describe the variables and instructions. They are notes by the author documenting the program so that the program is clear and easy to read. Comments are ignored by the computer.

In construction, before we can start, we must order our materials: “We need 500 large bricks, 80 half-size bricks, and 4 flagstones.” Similarly, in C, we must declare our variables before we can use them. We must name each one of our “bricks” and tell C what type of brick to use.
After our variables are defined, we can begin to use them. In construction, the basic structure is a room. By combining many rooms, we form a building. In C, the basic structure is a function. Functions can be combined to form a program.

An apprentice builder does not start out building the Empire State Building, but rather starts on a one-room house. In this chapter, we will concentrate on constructing simple one-function programs.

Part C Exercises

The basic elements of a program are the data declarations, functions, and comments.
A brief introduction to how these can be organized into a simple C program.
The basic C program is:

```
#include "stdio.h"
#include "conio.h"
main()
{
   int x,y,z,t;
   scanf("%d%d%d",&x,&y,&z);
   if (x>y)
       {t=x;x=y;y=t;} /* swap the values of x and y */
   if(x>z)
       {t=z;z=x;x=t;} /* swap the values of x and z */
   if(y>z)
       {t=y;y=z;z=t;} /* swap the values of z and y */
   printf("small to big: %d %d %d\n",x,y,z);
   getch();
}
```

Unit Four Automotive Engineering

Lesson 18 Principle of Automobile Engine Operation

教学目的和要求

本文介绍了汽油发动机的组成及工作原理。通过本文的学习，可以了解汽油发动机各总成及其重要部件的英文表达，进一步掌握四冲程发动机的工作过程。

重点和难点

（1）重点掌握点火系统、燃油供给系统各部件的英文名称及其工作原理。
（2）掌握四冲程发动机工作过程的英语表达。

Part A Text

Overall of Gasoline Engine

The engine, in conjunction with the chassis, electrical equipment, body, constitutes a traditional automobile. It is used to supply power for an automobile, which is called the heart of an automobile. The gasoline engine includes two mechanisms, which are crank mechanism and valve mechanism, and five systems, which are the starting system, ignition system, fuel system, lubricating system, and cooling system[1].

(1) Crank Mechanism.
The crank mechanism converts the reciprocating motion of the piston to the rotary motion of the crankshaft to drive the vehicle, which includes cylinder head cover, cylinder head, cylinder block, cylinder sleeve, cylinder gasket, oil pan, piston, connecting rod, crankshaft, and so on.

(2) Valve Mechanism.
It is used to open and close the valves called intake valves and exhaust valves at just the right time. When to open or close them is achieved by the camshaft rotating. Most engines have two camshafts, one called intake camshaft is necessary for opening and closing the intake valves, while the other called exhaust camshaft for opening and closing the exhaust valves[2]. In addition,

when the two camshafts are located in the cylinder head, the arrangement is called dual overhead-camshaft (DOHC), which is becoming main stream among automotive manufacturers. In order to improve the efficiency of intake and exhaust, there are generally two intake valves and two exhaust valves per cylinder. The time at which valves open and close and the duration of valve opening is called valve timing, which is stated in degrees of crankshaft rotation. Since different rotating speed of engine needs different requirement for valve timing, some engines equipped with variable valve timing(VVT) are developed to achieve the best valve timing for high speed and low speed.

(3)Starting System.

It mainly consists of the battery, starter, starting relay and the ignition switch. The battery provides electrical energy for starting, then once the engine is running, the alternator supplies all the electrical components of the vehicle. The starter is a DC electric motor designed to crank the engine.

(4) Ignition System.

It supplies electric sparks for the engine at the right time that are strong and that are needed to ignite the air-fuel mixture in the combustion chamber[3]. There are many different types of electronic ignition system. Most of these systems can be placed into one of two distinct groups: the distributor ignition system, and the distributor-less ignition system, which is also called direct ignition system (DIS), as shown in Fig.18.1.

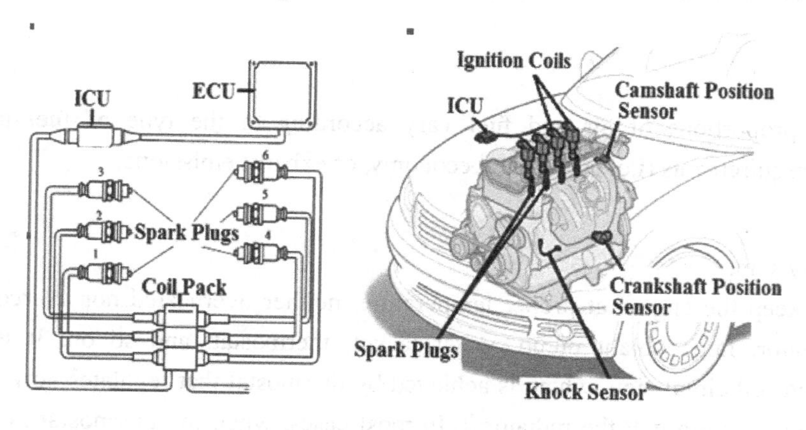

Fig.18.1 The Distributor-less Ignition System

The distributor-less ignition system whose spark plugs are fired directly from the ignition coils is becoming more and more popular. The major components of a distributor-less ignition system are: ECU, ignition control unit, magnetic triggering device, ignition coil packs, spark plugs. The ignition coil is an electromagnetic device that could convert low voltage produced by the battery into high voltage to puncture spark plug. The spark plugs are screwed into the combustion chambers in the cylinder head. The spark timing is controlled by the ICU and the ECU. The distributor-less ignition system uses either a magnetic crankshaft sensor, camshaft position sensor, or both, to determine crankshaft position and engine speed. This signal is sent to the ignition control module or engine control module which then energizes the appropriate coil.

The distributor-less ignition system may have one coil per cylinder, or one coil for each pair of cylinders. Coil- on-plug has individual ignition coil for each spark plug, which is installed directly on it, so that it makes the spark as high as possible and improves the reliability of system. Some using one ignition coil per two cylinders is often known as the waste spark distribution method[4].

(5) Fuel System.

Gasoline direct injection (GDI) is a type of EFI that sprays gasoline directly into the combustion chamber. The GDI injectors are mounted in the cylinder head and spray fuel directly into the combustion chamber instead of the intake port[5]. This system can be divided into three basic subsystems, which are the fuel delivery system, air supplying system, and the electronic control system.

The air supplying system includes air cleaner, air flow sensor, throttle valve, intake manifold and so on. The throttle valve controls how much air enters the cylinder which is achieved by the driver to depress the accelerator pedal. The fuel delivery system contains oil tank, electric fuel pump, fuel filter, oil pressure regulator, oil rail, fuel injector, and so on. Since the injection pressure is kept constant by the oil pressure regulator, the amount of fuel injecting from the fuel injector depends on injector pulse width that is decided by ECU from receiving signals by all kinds of sensors.

The relative proportions of air and fuel vary according to the type of fuel used and the performance requirements (i.e. power, fuel economy, or exhaust emissions).

(6) Cooling System.

It is used to keep the engine at proper temperature, neither overheated nor overcooling, which includes radiator, fan, coolant pump, water jacket, thermostat, and so on. It involves little circulation and full circulation, which is achieved by thermostat that regulates how much coolant is permitted to flow through the radiator[6]. In most cases, when the thermostat fails, it remains closed and the engine would overheat because coolant cannot flow into the radiator. Instead, if it remains open, the engine would overcool.

(7) Lubrication System.

The purpose of the lubrication system is to circulate clean oil which is at sufficient quantity, just right temperature through the friction surface of the engine. And it forms a film between the friction surface to achieve liquid friction, thus reducing the friction resistance, achieving lower power consumption, and reducing engine wear, with the purpose of improving engine reliability and durability.

Lubricating systems are divided into pressure lubrication, which is used for main bearing journals of crankshaft, shell bearing of connecting rod and main bearing journals of camshaft, and splash lubrication, that mainly lubricates pistons and cylinders wall, and so on. Oil pan, oil pump, oil filters, relief valve, and oil sensors mainly constitute lubrication system.

The Four-stroke Gasoline Engine Cycle

The four-stroke gasoline engine cycle consists of intake stroke, compression stroke, power stroke, and exhaust stroke, as shown in Fig.18.2.

(1) Intake Stroke.

On the intake stroke, the intake valve is open, and the exhaust valve is closed. The piston is moving downward from TDC to BDC.As this happens, the volume of the cylinder increases, the pressure in the cylinder drops. When the pressure falls below the atmospheric pressure, forming a certain degree of vacuum, so the air and fuel mixture is drawn from the open intake valve to the cylinder. Due to the resistance of the intake system, the pressure in the cylinder is about 0.075～0.09 MPa.

(2) Compression Stroke.

After the piston reaching BDC, it begins to move upward. At the same time the intake valve is also closed, so the cylinder is sealed. As the piston moving, the volume of the cylinder is reduced, so the mixture is compressed. The temperature and pressure of the mixture in cylinder increase simultaneously, which reaches up to fifteen bars and 800K respectively that outclasses the ignition temperature of gasoline (about 263K), so the mixture is easily ignited. Further uniform mixing of air and fuel mixture could get ready for full combustion.

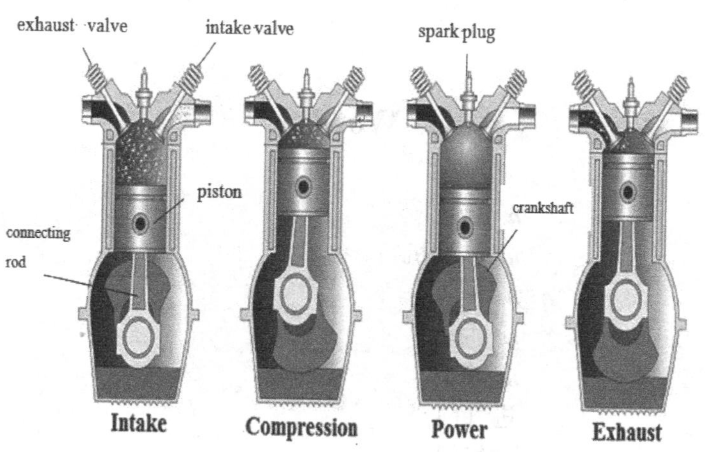

Fig.18.2 The Four Strokes

The ratio of the maximum volume of mixture in the cylinder before compression and the minimum volume after compression is called compression ratio. In other words, it is equal to the ratio of the swept volume and clearance volume to the clearance volume. The greater the compression ratio, the higher the pressure and temperature of the mixture at the end of the compression stroke, the faster the combustion rate, thus the greater the engine power output, the better fuel economy. But if the compression ratio is too large, it could not improve the combustion condition further, on the contrary, there will be some abnormal combustion phenomena, such as surface ignition and deflagration.

(3) Power Stroke.

Shortly before TDC is reached, the spark plug ignites the compressed air-fuel mixture and begins the ignition process. The angle of the crankshaft rotating from ignition time to reaching TDC is called ignition advance angle.

After the spark plug has ignited the compressed air-fuel mixture, the temperature and pressure in the cylinder increase quickly and force the piston downwards. The piston transfers power to the crankshaft by the connecting rod.

(4) Exhaust Stroke.

At the end of the power stroke, the exhaust valve has opened. The crankshaft pushes the piston from BDC to TDC through the connecting rod, then the combusted gases are expelled from exhaust valves by residual pressure of itself and thrust of piston. Exhausting gases fully could prepare for the next cycle to draw fresh air. When the exhaust stroke ends, the intake valve opens, it is the intake stroke again. Exhaust stroke and intake stroke are referred to as gas exchanging process.

To complete the full cycles it takes two revolutions of the crankshaft and one revolution of the camshaft.

Words

chassis	*n.* 底盘
gasoline	*n.* 汽油机
piston	*n.* 活塞
crankshaft	*n.* 曲轴
cylinder	*n.* 汽缸
intake valve	进气门
exhaust valve	排气门
DOHC	双顶置凸轮轴

combustion chamber	燃烧室
spark plug	火花塞
ignition coil	点火线圈
air flow sensor	空气流量计
throttle valve	节气门
oil pressure regulator	油压调节器
fuel injector	喷油器
thermostat	*n.* 节温器
TDC	上止点
BDC	下止点
compression ratio	压缩比
swept volume	汽缸工作容积
clearance volume	燃烧室容积
abnormal	*adj.* 反常的、异常的
deflagration	*n.* 爆燃
ignition advance angle	点火提前角

Notes

[1] The gasoline engine includes two mechanisms, which are crank mechanism and valve mechanism, and five systems, which are the starting system, ignition system, fuel system, lubricating system, and cooling system.

【注释】此句采用"分译法"进行翻译。

【译文】汽油发动机有两大机构和五大系统组成。两大机构分别是曲柄连杆机构和配气机构，五大系统分别是启动系统、点火系统、燃油供给系统、润滑系统和冷却系统。

[2] Most engines have two camshafts, one called intake camshaft is necessary for opening and closing the intake valves, while the other called exhaust camshaft for opening and closing the exhaust valves.

【注释】

① called intake camshaft 和 called exhaust camshaft：这是定语从句，省略 that is，修饰前面的 one 和 the other。

② while the other called exhaust camshaft for opening and closing the exhaust valves.

翻译该句时需要将省略的 is necessary 翻译出来。

③ 整句也采用"分译法"翻译。

[3] It supplies electric sparks for the engine at the right time that are strong and that are needed to ignite the air-fuel mixture in the combustion chamber.

【注释】本句为并列定语从句，that are strong and that are needed to ignite the air-fuel mixture in the combustion chamber 修饰先行词 electric sparks。

[4] Coil- on-plug has individual ignition coil for each spark plug, which is installed directly on it, so that it makes the spark as high as possible and improves the reliability of system. Some using one ignition coil per two cylinders is often known as the waste spark distribution method.

【注释】

① Coil-on-plug：独立点火，简称 COP，直译为“线圈在火花塞上”。线圈直接安装在火花塞上即一个汽缸有一个独立线圈，俗称“独立点火”。这种点火系统产生的火花更强，提高了点火系统的可靠性。

② waste spark distribution：双缸同时点火方式，可直译为“废火分配方式”，即利用一个点火线圈，对活塞接近压缩行程上止点和排气行程上止点的两个汽缸同时进行点火的高压配电方法。其中，活塞接近压缩行程上止点的汽缸的点火方式为有效点火，而活塞接近排气行程上止点的汽缸的点火方式为无效点火。

[5] The GDI injectors are mounted in the cylinder head and spray fuel directly into the combustion chamber instead of the intake port.

【注释】GDI 即汽油直接喷射式发动机，它的工作原理是先利用安装在缸盖上的喷油器将汽油直接喷到缸内，再利用缸内气流运动将燃料雾化，最终实现燃烧。

[6] It involves little circulation and full circulation, which is achieved by thermostat that regulates how much coolant is permitted to flow through the radiator.

【注释】本句为双重定语从句，which is achieved by thermostat 修饰 little circulation and full circulation，而 that regulates how much coolant is permitted to flow through the radiator 修饰 thermostat。

例句：This is a solution which is added to the cooling system to lower the temperature at which freezing of the coolant occurs.

该例句为双重定语从句，which is added to the cooling system to lower the temperature 修饰 a solution，which freezing of the coolant occurs 修饰 the temperature。

Part B Reading Materials

The engine can be classified in several ways depending on different design features:

(1) Number of Cylinders.

Current engine designs include three-cylinder, four-cylinder, six-cylinder, and even eight-cylinder, twelve-cylinder engine. Four-cylinder and six-cylinder are most common at present. Since the swept volume of all the cylinders is called engine capacity, the number of cylinders mainly depends on engine capacity requirement. Three-cylinder engine could satisfy the needs of energy saving and emission reduction, which is paid attention to by automobile manufacturers.

(2) Cylinder Arrangement.

The way that engine cylinders are arranged is called the engine configuration. An engine can be inline, flat (horizontal-opposed), or V-type. Other more complex designs have also been used, such as W8, W12, W16, W18, which are mainly used on Volkswagen.

In-line engines have the cylinders in a line, which is suitable for three-cylinder, and four-cylinder engine. With the number of the cylinders increasing to six, V-type is fitted for decreasing the length of engine. When the number of the cylinders increases further, W-type appears. W-type engine is a newly developed engine. W-type engine has a shorter crankshaft, but it is complicated in structure and expensive in cost, compared to a V-type engine.

Different Cylinder arrangements have different advantages and disadvantages in terms of smoothness, manufacturing cost and shape characteristics.

(3) Fuel Type.

The types of fuel are mainly divided into gasoline, diesel, CNG, and methanol. In China, gasoline is used on most passenger vehicle. As low emissions fuel, CNG serves for public transport. Because of venting dangerous gas, methanol develops slowly. With traditional energy prices rising and the world placing more and more emphasis on pollution reduction and environmental protection, governments, consumers and vehicle companies have turned to low zero emissions *green* vehicles, so this conventional fuel vehicle could be replaced gradually.

Part C Exercises

1. Fill in the blank, as shown in Fig.18.3.

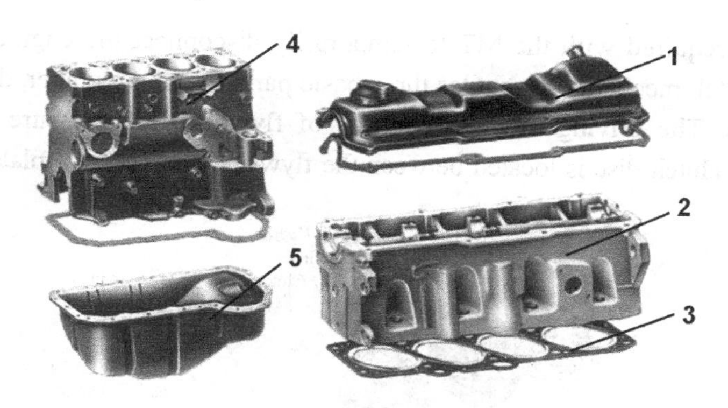

Fig.18.3 Engine Body

1. ____________ 2. ____________ 3. ____________
4. ____________ 5. ____________

2. Refer to the relevant information and introduce a W engine.

Lesson 19 An Introduction to Chassis

教学目的和要求

本文介绍汽车底盘的组成及各部分工作原理。通过本文的学习，可以了解底盘各系统的英文表达。

重点和难点

（1）掌握动力转向系统组成及其工作原理的英文表达。

（2）掌握制动系统工作原理及相关英文表达。

Part A Text

The chassis is composed of power train, steering system, suspension system, and braking system, which is used to support the body, accept the driving force generated by the engine, and ensure the vehicle driving normally[1].

Power Train

The power train could transfer the power from the engine to the driving wheels. Different engine torque and engine speed are achieved in accordance with the tractive power demand of the vehicle. The location of the engine and driving wheels determines whether the vehicle is classified as FF, FR, RR, MR, or nWD. The power train mainly includes clutch, transmission, final drive, differential, driving wheels and so on.

1) Clutch

The clutch is only required with the MT to temporarily disconnect the engine from the driving wheels. The dry clutch mechanism includes three basic parts, driving member, driven member and operating members. The driving member consists of flywheel and pressure plate. The driven member also called clutch disc is located between the flywheel and pressure plate (Fig.19.1). When

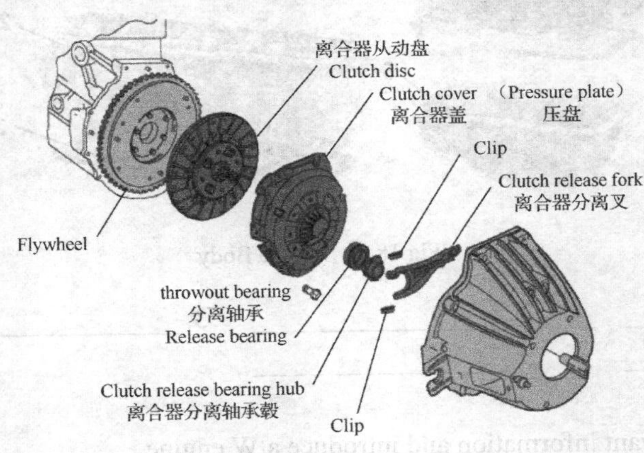

Fig.19.1 Components of Clutch

the clutch pedal is depressed, the clutch linkage operates the clutch fork, and moves the throwout bearing against the pressure plate release levers. For the diaphragm spring clutch, pushing center of diaphragm inwards causes outer edge to move away from flywheel, and it disengages the engine from the transmission. When the clutch pedal is released, the pressure plate forces the clutch disc against the flywheel.

When we will move the vehicle we should operate as follows:

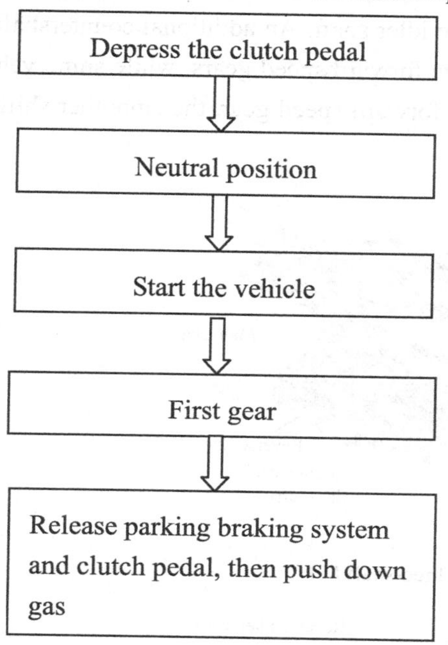

When we will change gears we should operate as follows:

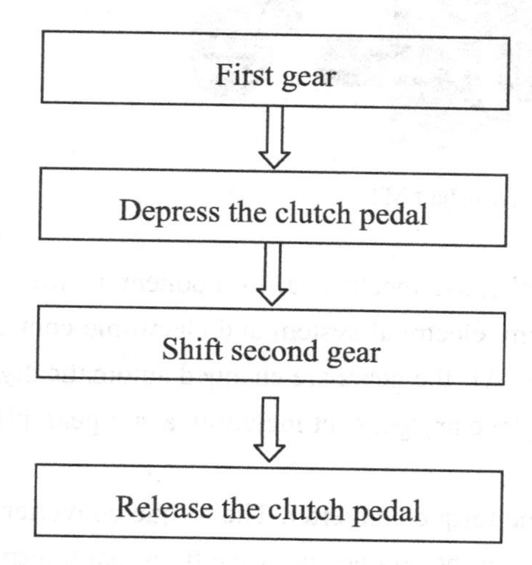

2) Transmission

The transmission is the key device that makes better use of engine power and torque. Thus, we can choose more speed and less torque (higher gear) or less speed and more torque (lower gear) to suit different driving conditions. There are mainly three types of transmission-manual (MT) transmission, automatic (AT) transmission, and continuously variable transmission (CVT).

(1) Manual Transmission (MT).

A manual transmission can be divided into three-shaft MT (Fig.19.2) and two-shaft MT (Fig.19.3). Two-shaft MT contains some shift forks, collars, different sizes of gears along the shafts such as input and output shaft, and reverse idler shaft. An additional countershaft has been added to the three-shaft MT. MT usually has five forward speed gears, while some vehicles now have as many as six and seven gears. The more forward speed gear, the smoother shifting is and more fuel-efficient the vehicle becomes.

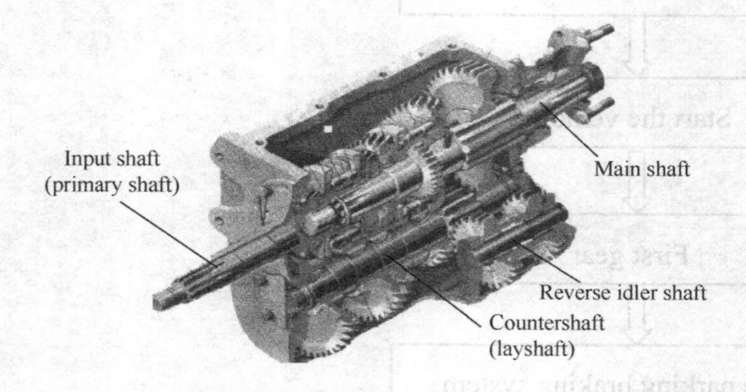

Fig.19.2 Three-shaft MT

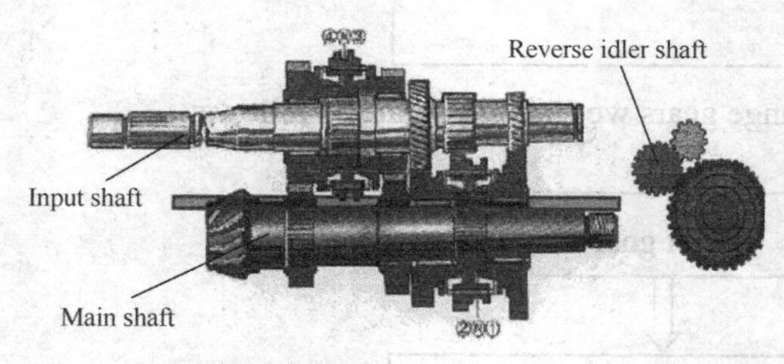

Fig.19.3 Two-shaft MT

(2) Automatic Transmission (AT).

An automatic transmission is the most complicated mechanical component in the vehicle. It consists of mechanical system, hydraulic system, electrical system and electronic control system, all working together in perfect harmony. For the AT, the gears are changed automatically, which is accomplished through the torque converter, a planetary gear set including a sun gear, planet gears and a carrier, and other shift operating devices.

The front section of the transmission houses the torque converter. The torque converter is similar to the clutch in the MT. It is the coupling between the engine and drive train that transmits power to the drive wheels. The planetary system works on the following principles: with one part held stationary, torque and power can be applied to a second member to make a third member rotate. In other words, with one of the three members held stationary, torque is applied to either of the other two members and output torque is delivered from the third member.

All automatic transmissions have a gearshift that has at least four positions: park (P), reverse (R), neutral (N) and drive (D).

Steering System

The steering system provides the driver with a means for controlling the direction of the vehicle as it moves. There are two types of steering system, which are manual steering system and power steering system. Manual steering system includes steering operating mechanism, steering box and steering linkage. The steering box is the heart of the steering system, which is used to convert the rotary motion of the steering wheel into straight line motion. There are many kinds of steering box, but two kinds of them are most popular, one is rack-and-pinion steering box which is used on many passenger vehicles (Fig.19.4), the other is recirculation ball steering box used on most commercial vehicles, especially on heavy commercial vehicles, as shown in Fig.19.5.

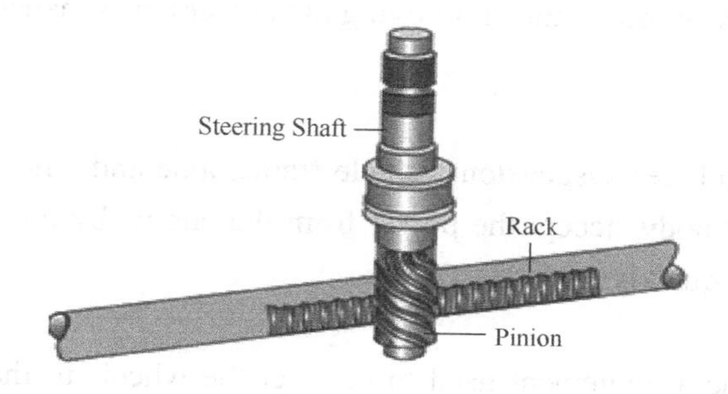

Fig.19.4 Rack-and-pinion Steering Box

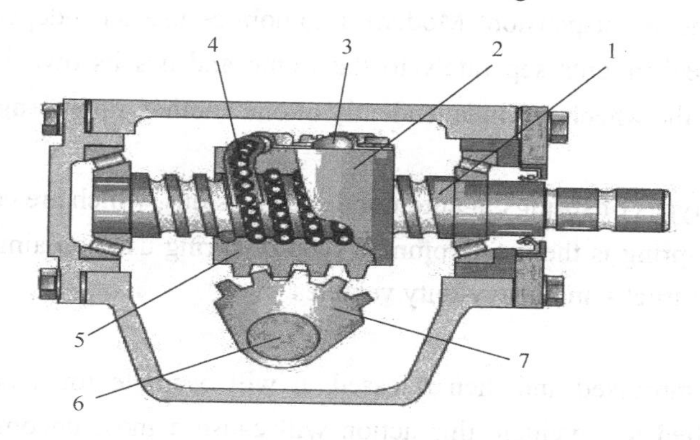

1. worm shaft 2. ball nut 3. ball guide 4. balls 5. rack 6. sector shaft 7. sector gear

Fig.19.5 Recirculation Ball Steering Box

Power steering system adds a series of assisting power mechanism to reduce the driver effort. Depending on the type of power, it can mainly be classified into hydraulic power assisted steering system (HPAS) and electric power assisted steering system (EPAS).

HPAS consists of an oil reservoir, a steering pump, a steering control valve, a steering cylinder, and corresponding pipes. The steering control valve provides an oil pressure which corresponds to the rotary motion of the steering wheel for the steering cylinder. The steering cylinder converts the applied oil pressure into an assisting force which acts on the steering box and which intensifies the steering force exerted by the driver. When the steering wheel is centered, the steering control valve is not actuated, the oil delivered by the steering pump all flows back to the oil reservoir.

EPAS is designed to use an electric motor to provide directional control of the vehicle. It mainly contains torque sensor, speed sensor, ECU and electric motor. When the driver turns the steering wheel, the torque sensors sense the steering direction and the amount of torque and in addition the speed sensor measures the speed of vehicle. And then the two signals are fed to ECU. ECU commands the electric motor to provide the power-assisted steering according to the two signals. The EPAS is replacing the HPAS and is destined to become mainstream among automotive manufacturers. Since EPAS does not require engine power to operate, a vehicle equipped with an EPAS may reduce fuel consumption, thus improving the power performance of automobile.

Suspension System

The suspension system includes suspension, vehicle frame, axle and wheels. It's used to flexibly connect the wheels to the body, accept the power from the engine by the power train, and damp out the spring oscillations quickly.

The suspension covers the arrangement used to connect the wheels to the body which includes spring element, shock absorber and guide mechanism. It can be classified into independent suspension and dependent suspension. Modern automobiles use an independent suspension. In this system, each wheel mounts separately to the frame and has its own individual spring and shock absorber. Thus, the wheels act independently of one another, improving ride comfort.

There are three basic types of springs in automobile suspensions, which are coil spring, leaf spring and torsion bar. Coil spring is the most common type of spring used on almost all vehicles. Leaf spring is used on most trucks and heavy-duty vehicles.

When a spring is compressed and then released, it will oscillate for a period of time before coming to rest. Applied to a vehicle this action will cause a most uncomfortable ride. So the vehicle is equipped with a shock absorber to absorb the energy stored in the spring and reduce the time that the vehicle is bouncing. Hydraulic shock absorber is widely used in automobile suspension system.

Four-wheel alignment is necessary on today's design vehicles. Correct alignment is vital to vehicle control, not only for safety but also for comfort when driving a vehicle. Four-wheel aligner is a precise measuring instrument that is used to measure the wheel alignment parameters and compare them with the specifications provided by vehicle manufacturer[2]. It also gives instructions to the user for performing corresponding adjustments so as to get the best steering performance and reduce tire wear. The four-wheel aligner can be used to measure the most wheel alignment parameters, such as toe-in, positive camber, caster, kingpin angle, thrust angle etc.

Braking System

Braking system is the most important safety system on an automobile. The braking system on a vehicle has three main functions: it must be able to make the driving vehicle forcibly decelerate or even stop according to the driver's requirements, it must be able to stabilize stopped vehicles under various road conditions, including on ramps, and it must be able to keep the speed of the vehicle stably when going downhill. Typically, each automobile has two completely independent braking system according to the different functions. One is service braking system which is normally foot-operated, and the other is parking braking system with hand-operated.

With the constant development of automobile electrical technology, the parking braking system has been changed from the primary mechanical type into electrical type gradually. Electric parking braking system is able to release before driving or strain after stopping the engine automatically, and cuts the difficulties and dangerous rates of forgetting to release or sliding on slope[3].

A service braking system consists of an energy-supplying device, a control device, a transmission device and the brake. The braking action is achieved as a result of the friction developed by forcing a stationary surface into contact with a rotating surface (the drum or disc). A brake is a device which inhibits motion, decisive part of a vehicle. Each wheel has a brake assembly, of either the drum type or the disc type, hydraulically operated or pneumatic operated when the driver depresses brake pedal.

Drum brake has two shoes, anchored to a stationary back-plate, which are internally expanded to contact the drum by hydraulic cylinders. The structure of the drum brake is as shown in the Fig.19.6. When the brake pedal is depressed by the driver, the two pistons of the master cylinder move in the two sections, which force brake fluid out and through the brake lines to the wheel cylinder. This causes the brake shoes to move into contact with the brake drum, applying friction to the brake drum, forcing the wheel to slow or stop. When the brake pedal is released by the driver, return springs on the brake shoes contract and pull the shoes away from the braking surface. The movement of the wheel cylinder pistons forces fluid back to the master cylinder to replenish the reservoir.

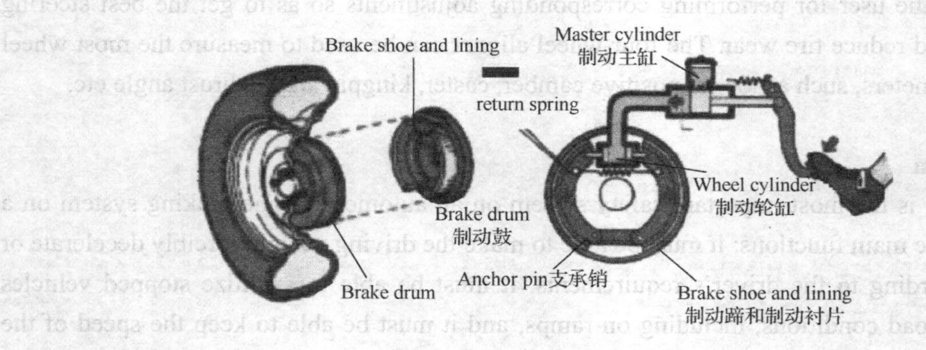

Fig.19.6 The Structure of Drum Brake

Hydraulically actuated disc brakes are the most commonly used form of brake for passenger vehicles. There are two kinds of disc brakes currently used in modern automobiles: one is a disc brake with fixed caliper, and the other is disc brake with floating caliper (Fig.19.7). The disc brake mainly consists of: brake pad, brake caliper, brake disc. Brake caliper contains two brake pads with its friction surfaces facing the rotor. Brake disc (brake rotor in US English), is usually made of cast iron. Exposed to the air, disc brakes radiate the heat to the air better than drum brakes. This means that the brake can be operated continuously for a longer period, in other words, they have a greater resistance to heat fade. Disc brakes as used in most passenger vehicles are slowly beginning to be used in commercial vehicles as well.

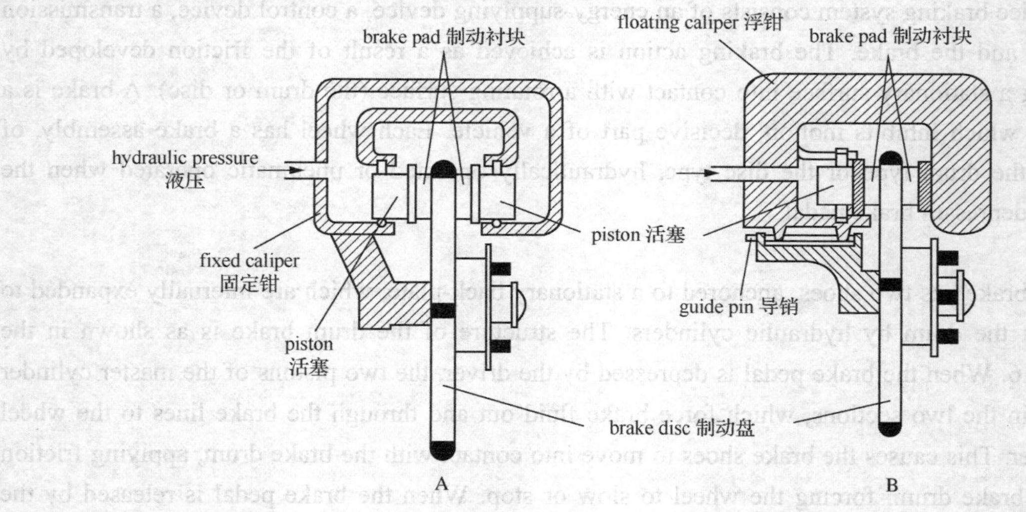

Fig.19.7 The Structure of Disc Brake

Words

chassis	*n.* 底盘
power train	传动系统
steering system	转向系统

suspension system	行驶系统
braking system	制动系统
diaphragm spring clutch	膜片弹簧离合器
clutch pedal	离合器踏板
neutral	*n.* 空挡
MT	手动变速器
AT	自动变速器
CVT	无极变速器
torque converter	液力变矩器
steering box	转向器
HPAS	液压助力转向系统
steering control valve	转向控制阀
steering cylinder	转向液压缸
EPAS	电控助力转向系统
torque sensor	扭矩传感器
independent suspension	独立悬架
coil spring	螺旋弹簧
leaf spring	钢板弹簧
torsion bar	扭杆弹簧
shock absorber	减振器
four-wheel alignment	四轮定位
electric parking braking	电子驻车制动系统

Notes

[1] The chassis is composed of power train, steering system, suspension system, and braking system, which is used to support the body, accept the driving force generated by the engine, and ensure the vehicle driving normal.

【译文】汽车底盘由传动系统、行驶系统、转向系统和制动系统四部分组成，其作用是支撑车身，接受发动机的动力，确保汽车正常行驶。

【注释】Which 引导的定语从句修饰 the chassis。

[2] Four-wheel aligner is a precise measuring instrument that is used to measure the wheel alignment parameters and compare them with the specifications provided by vehicle manufacturer.

【译文】四轮定位仪是一种用来检测车轮定位参数并将其与汽车制造商提供的说明书里的标准值进行比较的精确测量装置。

【注释】that is used to measure the wheel alignment parameters and compare them with the specifications provided by vehicle manufacturer 作为定语从句，修饰 a precise measuring instrument，而 provided by vehicle manufacturer 又作为定语修饰 the specifications，省略 that is。此句为双重定语从句。

[3] Electric parking braking system is able to release before driving or strain after stopping the engine automatically, and cuts the difficulties and dangerous rates of forgetting to release or sliding on slope.

【译文】电子驻车制动系统可在起步前自动释放或在熄火后自动拉紧，省去了忘记解除驻车制动或坡道起步溜车等情况的困扰与风险。

【注释】电子驻车系统的工作原理与机械式驻车制动相同，均通过制动盘与制动片产生的摩擦力达到制动目的，只不过控制方式从之前的机械式驻车制动拉杆变成了电子按钮。

Part B　Reading Materials

Anti-lock braking system (ABS) is used to prevent the wheels from locking under hard braking (Fig.19.8). It includes wheel speed sensors, an electronic controller and pressure modulation valves. The wheel-speed sensor monitors the motion of the wheel. If one wheel shows signs of locking which exceed defined critical values, the electronic controller sends commands to the solenoid-valve unit to stop or reduce the buildup of wheel-brake pressure until the danger of lock-up has passed. The brake pressure must then be built up again in order to ensure that the wheel is braked normally.

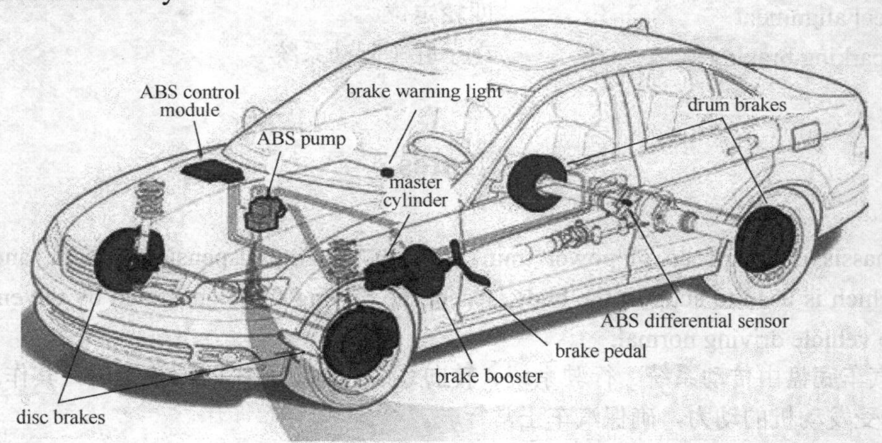

Fig.19.8　ABS

As the subsystem of ABS, EBD which is abbreviation of Electronic Brake-force Distribution is designed to automatically modulate the brake force on the front wheels and the rear wheels according to the load of the vehicle , the condition of the road and the adhesion between the tire sand road surface in order to get more balanced and closer to the ideal brake force distribution.

The ESC is a new type of active safety system for vehicles that is a further expansion of the functions of Anti-lock braking systems (ABS) and Traction Control System (TRAC) and that uses sensors to detect when a driver is about to lose control of the vehicle and automatically intervenes to provide stability and help the driver stay on the intended course, especially in over-steering and under-steering situations.

Industry experts have hailed electronic stability control as a milestone in automotive safety, comparing it to seatbelts and SRS. ESC helps to prevent from rollovers and skids when the vehicles drive on the road, thus as the important safety choice for consumers when purchasing a new vehicle.

Part C Exercises

(1) What are the benefits that EPAS compares to HPAS?

(2) What are the advantages of disc brake in comparison to drum brake?

Lesson 20 Automobile Inspection and Maintenance

教学目的和要求

本文介绍使用汽车故障诊断仪对发动机进行诊断的一般步骤，以发动机不能正常启动为例，分析了引起该故障可能的原因，并逐一进行检测和诊断，使读者掌握汽车检测和诊断方法的相关专业术语的英文表达。

重点和难点

（1）掌握读取故障代码和清除故障代码的一般方法及其英语表达。

（2）掌握在没有故障代码的情况下，检测和诊断故障的一般方法。

Part A Text

Along with the application of advanced electronic control technology and automatic failure diagnosis to automobiles, the importance of decoder application has become more and more evident in automobile failure diagnosis. The application level of automatic failure diagnosis technology mainly depends on the application level of decoders. Wide extension of the advanced and specialized decoders, and application level promotion of domestic decoders are both required. X431 gx3 is a newly developed automobile diagnostic computer, the most advanced automobile diagnostic technology[1], as shown in the Fig.20.1.

Fig.20.1 X431 gx3

There are 10,000 different diagnoses the vehicle can give, so putting your finger on the right one can be difficult. Figuring out how to fix the problem is another kettle of fish, but the first step is to know what the vehicle is trying to tell us. The language it speaks is OBD, which stands for On Board Diagnostics.

Reading the fault code and clearing the code with the fault diagnosis instrument should follow the following procedure:

Turn off the ignition switch after confirming the fault code by the fault indicator light. Connect the fault diagnosis instrument to the fault diagnosis interface on the lower side of the dashboard on the driver's seat side. Then turn on the ignition switch to read the fault code according to the operation requirements of the fault diagnostic instrument, HFM Diagnostic as shown in the Fig.20.2. Fault code is still kept in the computer after clearing of fault, the code should be cleared. Press the exit key to return to the upper grade catalogue, select *clear fault code*, press confirmation key to clear the code. Turn the ignition switch to the off position and disconnect the fault diagnosis device from the fault diagnosis interface.

HFM Diagnostic
1. Data stream
2. Read stored code
3. Read current code
4. clear fault code

Fig.20.2 HFM Diagnostic

Note: Do not disconnect the battery cable before reading the fault code. After the power off, the fault code will be eliminated automatically, which will affect the right judgment for the fault. So it is necessary to read the current fault code before considering the power outage. In addition, when disconnecting an automobile battery you may also run into trouble with theft deterrent radios and even factory installed car alarm systems. If we cleared the code by disconnecting the battery and there is a hard failure in the engine control or emissions system then the light will just pop right back on.

When the automotive electronic control engine is not working properly, and the self diagnostic system does not have fault code output, it is necessary to rely on the operator's inspection and judgment to determine the nature of the fault and the location of the fault.

Engine Cranks but Will not Start

A dead electrical fuel pump can prevent an engine from starting, but so can a problem with the ignition system or low compression in the cylinder[2]. When this failure occurs, normally, the first thing to check would be the spark. This can be done by connecting a spark plug tester to a plug wire while the engine is cranking. The tester must be grounded to the engine block for a good electrical connection.

If an engine has a coil-on-plug ignition system, and it is possible to remove one of the ignition coils, do so and place a spark plug in the end of the coil. Then place the ignition coil and spark plug so the spark plug is touching metal on the engine. If the ignition system is working properly, you should see a series of sparks while cranking the engine. No spark would indicate an ignition problem such as a bad crankshaft position sensor, ignition module or ignition coil.

If the ignition system is normal, that leaves a lack of fuel as the most likely cause of it. But is it the fuel pump or something else? If the electric fuel pump does not deliver adequate fuel pressure and volume to the engine, the engine may not start or run properly. Low fuel pressure can cause hard starting, a rough idle, misfiring. An electric fuel pump that can deliver adequate pressure but not enough volume may allow the engine to start and idle normally, but it will starve the engine for fuel and cause a loss of power when the engine is under load, accelerating hard or cruising at highway speeds.

Possible causes of a fuel-related no start include:

(1) A dead electric fuel pump (could be the electromotor, pump relay or a fault in the pump wiring circuit), its structure as shown in the Fig.20.3.

(2) A plugged fuel filter.

(3) Low fuel pressure (weak pump, restricted fuel line, low voltage to the pump or a defective fuel pressure regulator).

Fig.20.3 The Structure of Electrical Fuel Pump

If the electric fuel pump runs and generates normal pressure to the engine, but the engine still does not start, the problem may be:

(1) No voltage to the fuel injectors(the injector wire connector plugs are not plugged in)

(2) No pulse signal to the injectors from the PCM (no crank or cam sensor input to the PCM, or a bad driver circuit in the PCM, or a wiring harness problem)

(3) A shorted fuel injector.

Measuring Fuel Pressure

Depending on the application, the fuel system may require anywhere from 30 to 80 psi of fuel pressure to start and run. NOTE: Electrical fuel injection engines are very sensitive to fuel pressure. If pressure at the engine fuel rail is even a couple pounds less than specifications, the engine may not start or run well, or experience stalling problems. When there is too much fuel pressure, the engine runs rich. This causes an increase in fuel consumption and carbon monoxide (CO) emissions. An engine that is running really rich also may experience a rough idle, and possibly even carbon-fouled spark plugs.

A fuel pressure gauge is needed to check different fuel pressure, including static pressure, running pressure, residual fuel pressure, maximum pressure. Different vehicle manufacturers recommend different test procedures. On many European EFI systems, the OEMs recommend using a static pressure test with the engine and ignition off, which is done by bypassing the fuel pump relay and energizing the pump directly. While most domestic and Asian vehicle manufacturers provide a test fitting on the fuel rail so pressure can be checked with the engine running. If a vehicle does not have a pressure test fitting, the pressure gauge should be fitted into the fuel line just ahead of the injector fuel rail. Caution: Before hooking up pressure gauge, relieve all pressure in the fuel system.

1) Static Fuel Pressure Test

With the key on, engine off (with the fuel pump energized), fuel pressure should come up quickly and hold steady at a fixed value. Compare the pressure reading to specifications. If you get no pressure reading, check for voltage at the pump. If there is voltage but the pump is not running, it demonstrates that the pump is wrong. If the pressure reading is lower than normal, the cause may be a weak pump, a blockage in the fuel line or filter, or a faulty pressure regulator. Also, low voltage at the pump may prevent it from spinning fast enough to build up normal pressure.

2) Running Fuel Pressure Test

With the engine idling, compare the gauge reading to specifications. Fuel pressure should be within the acceptable range given by the vehicle manufacturer. If low, the problem may be a weak electric fuel pump, low voltage to the electric fuel pump, a clogged fuel filter, line or inlet sock inside the fuel tank, a bad fuel pressure regulator, or nearly empty fuel tank.

3) Residual Fuel Pressure Test

When the electric fuel pump is turned off or stops running, the system should hold residual pressure for several minutes. If pressure drops quickly, the vehicle may have a leaky fuel line, a leaky electric fuel pump check valve, a leaky fuel pressure regulator or one or more leaky fuel injectors. Low residual fuel pressure can cause hard starting.

4) The Maximum Output Pressure of the Electrical Fuel Pump Test

With the return line pinched shut, the pump should produce two times its normal operating pressure at idle. If the pressure rating does not go up with the return line blocked, the pump may not be able to deliver enough fuel at higher engine speeds.

Fuel Volume Test

An electrical fuel pump that delivers normal pressure may still cause drive ability problems if it can't deliver enough fuel volume to meet the engine's needs. A fuel volume test measures the volume of fuel delivered over a specified interval, which may therefore be the best way to evaluate the pump's condition. This test can be done by connecting a fuel flow gauge into the fuel supply line (Fig.20.4), or by disconnecting the fuel return line from the fuel pressure regulator and connecting a hose from the regulator to a large container. With the engine off, energize the pump and measure the volume of fuel delivered during the specified interval of time. As a rule, a good pump should deliver about 3/4 to one quart of fuel in 30 seconds. Causes of low fuel volume delivered include a worn fuel pump, a plugged fuel filter or inlet sock in the tank, obstructed fuel line or nearly empty tank.

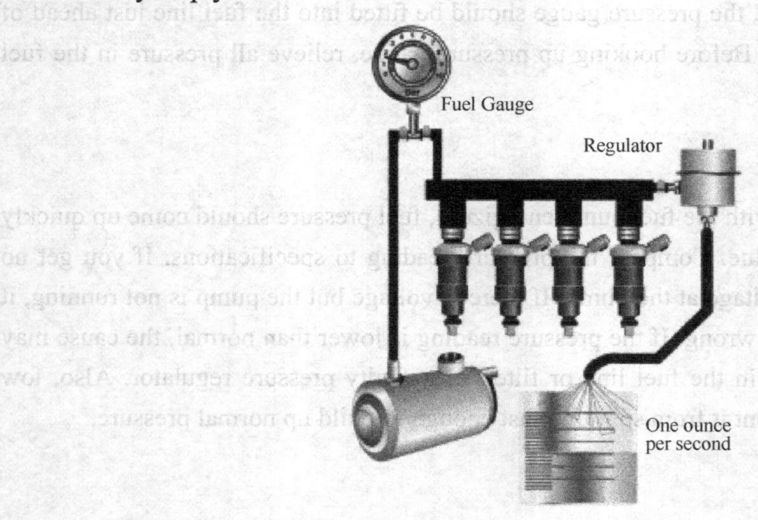

Fig.20.4 Fuel Volume Test

If the electrical fuel pump is running and delivering normal pressure to the engine, but will not start, the fuel injectors are probably not opening. We can measure the resistance of each injector with an ohmmeter. Unplug the injector and measure the resistance between the two terminals. If resistance is outside specifications (high or low), replace the injector.

If the fuel system is normal, test the compression pressure for each cylinder. Low compression pressure may be the fault of the valves or the pistons.

Words

OBD	车载自诊断
ignition switch	点火开关
EFI	电控燃油喷油系统
idle	*n.* 怠速
fuel pressure gauge	油压表
static fuel pressure	静态燃油压力
residual fuel pressure	保持压力
electric fuel pump check valve	电动燃油泵单向阀

Notes

[1] X431 gx3 is a newly developed automobile diagnostic computer, the most advanced automobile diagnostic technology.

【译文】X431 gx3 是最新开发的汽车诊断仪，代表了目前最先进的汽车诊断技术。

【注释】它是基于计算机应用程序推进的开放式诊断平台技术而研制的。开放式诊断平台代表了汽车诊断技术的最高水平，是未来汽车诊断的发展趋势。产品具有如下功能、开放汽车诊断功能、互联网更新功能、个人数字助理功能、多语言显示功能和打印功能。

[2] A dead electrical fuel pump can prevent an engine from starting, but so can a problem with the ignition system or low compression in the cylinder.

【译文】电动燃油泵出现故障能够导致发动机不能正常启动，点火系统出现故障或汽缸内压力低，同样也会导致发动机不能正常启动。

【注释】在启动系统正常工作的情况下，发动机能够正常起动的前提是空燃比正确、有火(且点火正时)、汽缸内密封性良好。

Part B Reading Materials

Most vehicles on the road are equipped with Onboard Diagnostic. OBD-II introduced in the mid-nineties, provides almost complete engine control and also monitors parts of the chassis, body and accessory devices, as well as the diagnostic control network of the vehicle. The OBD monitors all components of the engine management system. It can detect a malfunction or deterioration of the components even before the driver becomes aware of the problem. When a problem is detected that could cause an increase in air emissions, the OBD system turns on a dashboard warning light to alert the driver of the need to have a vehicle checked by a repair technician.

When the OBD system determines that a problem exists, a corresponding "Diagnostic Trouble Code" (DTC) is stored in the vehicles computer memory. The computer illuminates a dashboard light indicating "Service Engine Soon" or "Check Engine" or displays an engine symbol. This light serves to inform the driver that the computer has detected a problem and the vehicle needs service. By federal law this light can only be used to indicate an actual problem. The light cannot be used as a reminder for regular maintenance.

After fixing the problem, the service technician will turn off the dashboard light. There are some situations under which the vehicles OBD system can turn off the light automatically, if the conditions that caused the problem are no longer present. For example, if the OBD system evaluates a component or system, three consecutive times and no longer detects the initial problem, the dashboard light will turn off automatically. As a result, drivers may see the dashboard light turn on and then turn off.

Only qualified trained technicians equipped with appropriate diagnostic and repair equipment should conduct OBD related repairs. With the population of modern technology automobile growing, all dealerships and independent repair shops should have qualified personnel for this service. Vehicle owners should ask service facilities if they have the necessary training and equipment.

Part C Exercises

1. Analysis the reason of poor engine power performance, and introduce how to inspect.

Lesson 21 New Energy Vehicle

教学目的和要求

本文介绍新能源汽车的发展、分类及其优缺点，通过本文的学习，读者可以了解新能源汽车相关专业术语的英文表达。

重点和难点

（1）新能源汽车分类的相关英语表达。

（2）混合动力电动汽车的类型及插电式混合动力电动汽车的相关英语表达。

（3）燃料电池电动车组成及其工作原理的相关英语表达。

Part A Text

With the global energy crisis increasingly and more countries emphasis the environmental issues, the development of new energy vehicle has become an inevitable trend in the global automotive industry. New energy vehicles, with benefits of low power consumption, zero emission and non-pollution, fit well into this round of industrial revolution. The traditional automobile industry in China lags behind the foreign enterprises by two or three decades. Most key technologies are owned by foreign companies and it is difficult for the traditional auto companies to have any breakthrough. But after years of implementation of NEV, it is a rare opportunity that China's auto industry can catch up with the world auto industry. It could be the right time for the wise Chinese people to grasp this opportunity of the golden age, with scientific implementation, to develop it into a new form of nationwide industry. New energy vehicles are expected to be a key area in building China from a "big" auto power to a "strong" one. Starting in 2009, China's new energy auto industry experienced a robust expansion and it has become the world's largest market since 2015, according to a statement from the Ministry of Industry and Information Technology (MIIT).

With NEVs growing increasingly popular in China, the competition among electric motor companies becomes more and more intense. Since China has been the largest auto market and also the biggest NEV market worldwide, there is a great potential in China's electric motor market. In recent years, China's NEV industry has seen such a booming development that the potential policy change has received a lot of attention. The government has made a series of policies to promote the development of NEV market, such as subsidies and free vehicle license. With the development of new energy market, China has started to phase out NEV subsidies since 2016. It's obvious that the new energy subsidy has been shrinking. Moreover, the government will raise the bar of subsidy approval[1].

Under such a condition, enterprises have to rely on themselves independently in the future. In other words, only competent enterprises can survive the tough situation, while some may end with failure. It's reasonable to believe that many auto companies will make corresponding adjustment to their strategies after the release of new policies about NEV. This move has stimulated carmakers to reduce the purchasing costs of parts and promote technology development. After 2020, the expected costs of passenger vehicles' electric motors will be extremely low and may even lower than that of the same products sold in China, according to the demands of foreign carmakers. To respond the market changes, the NEV electric motor companies will prefer to manufacture one single product in a large quantity, so that the cost can be reduced, compared with the former mode of manufacturing products in various kinds and small quantity. More important, the emerging vehicle renting and sharing platforms will provide more opportunities as well as competition possibilities for NEV industry, which will attract more efforts and dedication of various auto companies into NEVs.

In the Chinese context, NEVs include battery electric vehicles (BEVs), hybrid electric vehicles (HEVs), especially plug-in hybrid electric vehicles (PHEVs), and fuel cell vehicles (FCVs). However, the definition sometimes covers broader vehicle technologies, such as alternative fuel vehicles. China witnessed a boom of electric vehicle investment in the past few years largely thanks to government's incentives. In terms of charging infrastructure, China built over 100,000 public charging poles in 2016, ten times the figure in 2015. A comprehensive charging grid has taken shape in big cities like Beijing, Shanghai, and Shenzhen.

1) Battery Electric Vehicle (BEV)

BEV is powered solely by the electricity stored in battery packs, and therefore does not require ICE or any other fuel source. The on-board battery used in these vehicles is of three types: Ni-MH, lead-acid, and Li-ion, out of which Li-ion is the most commonly used one. The chemical energy stored in rechargeable battery packs installed in BEVs propels the electric motors.

2) Hybrid Electric Vehicle (HEV)

According to the structures, HEV is divided into three types: series hybrid, parallel hybrid and series-parallel hybrid .

In a series hybrid vehicle, the engine and the electric motor are connected together in a series. The engine converts the fuel into the mechanical energy to drive the generator, generating electric power. Then the generator delivers the electric power through the inverter to the electric motor which moves the vehicle. There is no mechanical connection between the engine and the drive wheels, so the engine never directly drives vehicles.

In a parallel hybrid vehicle, the fuel tank supplies the fuel to the engine and the battery supplies the electric power to the electric motor. Both of the engine and the electric motor connect to the transmission in parallel.

Series-parallel hybrid combines the series hybrid system with the parallel hybrid system in order to make full use of the benefits of both systems.

All the three types of HEV include the regenerative braking feature. The regenerative braking is an energy recovery mechanism which converts wasted energy from braking into electricity and stores it in the battery. In the regenerative braking, the electric motor reverses direction so that the rotating vehicle wheels can turn it, producing electricity to store in the battery just like a generator.

3) Plug-in Hybrid Electric Vehicle

A Plug-in Hybrid Electric Vehicle (PHEV or PHV), also known as a plug-in hybrid, is a hybrid electric vehicle with rechargeable batteries that can be restored to full charge by connecting a plug to an external electric power source. A plug-in hybrid shares the characteristics of both a conventional hybrid electric vehicle and an all-electric vehicle: it uses a gasoline engine and an electric motor for propulsion, and a PHEV has a larger battery pack that can be recharged, allowing operation in all-electric mode until the battery is depleted.

PHEVs are full hybrid vehicles with a high power rated battery pack as in pure EVs, thereby, propelling the vehicle either through ICE or battery. Hence, these vehicles have a longer range and can use 100% electricity for propulsion. They are widely used in longer range and heavy-duty applications. Since the batteries of PHEVs can be recharged through a plug-in to external supply, they fall under the NEV(neighborhood Electric Vehicle) segment.

Plug-in hybrid electric vehicle (PHEV) will continue to be the mainstream in global private vehicle market and thus the government policy should encourage the promotion of PHEVs. The PHEV system has taken up nearly half of the global new energy vehicle (NEV) market shares. In China NEV market, PHEV sales took up 47 percent of the total NEV deliveries to private owners, excluding the deliveries of A00-class vehicle, whose proportion was so small in Europe or American NEV sales.

4) FCV

FCEV also called FCV (Fuel Cell Vehicle) is considered as the next generation electric vehicle. A fuel cell is a relatively new technology with a future potential market in automobile applications, such as passenger vehicles, commercial vehicles, and forklifts. Owing to the zero emission of GHG, major OEMs have shifted their focus toward the development of HFCVs. Therefore, this technology is expected to become commercially viable during the forecast period.

FCV combines hydrogen and oxygen to produce electricity used to move the vehicle, which can achieve long driving range. Unlike BEV, FCEV could be quickly refueled only for a couple of minutes at the filling station just like conventional ICE.

A typical FCEV is equipped with several main components (Fig.21.1), which mainly includes hydrogen tank, fuel cell stack, electric motor, power control unit, and so on. The hydrogen from the tank and oxygen from the air are pulled into the FC stack, which are converted into electricity to power the electric motor through a chemical reaction. FCV can drive more quietly, smoothly, and efficiently than other kinds of vehicle, only producing water vapor emitted from tailpipe without emission pollutants.

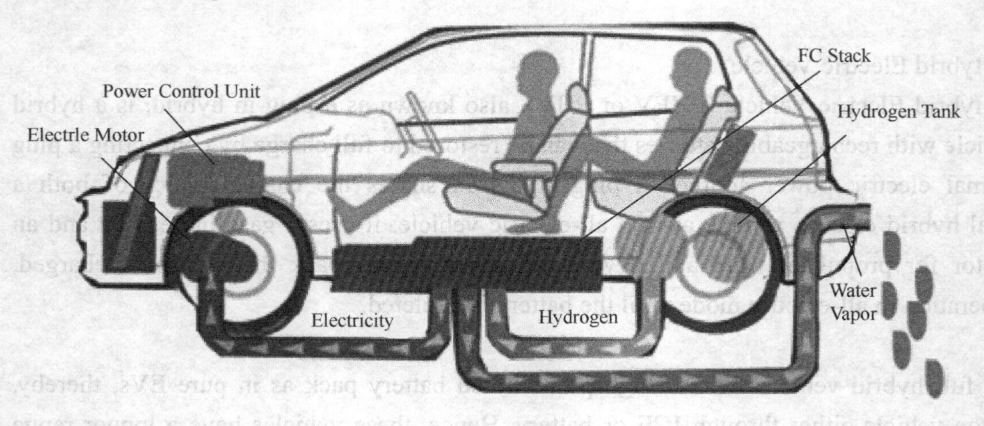

Fig.21.1 The Components of FCV

Electric drive vehicles can contribute significantly to lessen the dependence of the transport sector on imported oil as well as contributing to the development of a more resilient energy supply. NEV also has many other advantages, such as lower operating and maintenance costs, lower air pollution and greenhouse gas emissions, vehicle-to-grid.

Internal combustion engines are relatively inefficient at converting on-board fuel energy to propulsion as most of the energy is wasted as heat, and the rest while the engine is idling. While electric motors, on the other hand, are more efficient at converting stored energy into driving a vehicle. Electric drive vehicles do not consume energy while at rest or coasting, and modern plug-in cars can capture and reuse as much as one fifth of the energy normally lost during braking through regenerative braking.

All-electric and plug-in hybrid vehicles also have lower maintenance costs as compared to internal combustion vehicles, since electronic systems break down much less often than the mechanical systems in conventional vehicles, and the fewer mechanical systems on board last longer due to the better use of the electric engine. BEVs do not require oil changes and other routine maintenance checks.

Electric vehicles,as well as plug-in hybrids operating in all-electric mode, emit no harmful tailpipe pollutants from the onboard source of power, such as particulates, hydrocarbons, carbon monoxide,ozone, lead, and various oxides of nitrogen.

Plug-in electric vehicles offer users the opportunity to sell electricity stored in their batteries back to the power grid, thereby helping utilities to operate more efficiently in the management of their demand peaks. A vehicle-to-grid (V2G) system would take advantage of the fact that most vehicles are parked an average of 95 percent of the time. During such idle times the electricity stored in the batteries could be transferred from the PHEV to the power lines and back to the grid[2]. In the U.S. this transfer back to the grid have an estimated value to the utilities of up to $4,000 per year per automobile.

Research and development on NEV has spanned the past two decades, but because of cost, charging problems and other complicating factors, the development of the NEV industry has been significantly restrained. To further promote the healthy and sustainable development of the industry, more efforts should be made in improving the innovation system, advancing industrial transformation and upgrading, and strengthening the application of NEVs.

Words

NEV	新能源汽车
subsidy	*n.* 补贴
bar	*n.* 障碍
tough	*adj.* 艰苦的，困难的
rent	*v.* 出租
dedication	*n.* 奉献
BEV	纯电动汽车
HEV	混合动力电动汽车
PHEV	插电式混合动力电动汽车
FCV	燃料电池电动汽车
incentive	*n.* 刺激，鼓励，动机，诱因
infrastructure	*n.* 基础设施
pack	*n.* 一组，一副，一群
Ni-MH	镍氢电池
lead-acid	铅酸电池
Li-ion	锂离子电池
inverter	*n.* 逆变器
deplete	*v.* 耗尽，用尽

GHG	温室效应气体
OEM	原始设备制造商
resilient	*adj.* 有弹性的
sustainable	*adj.* 可持续的，可以忍受的，可以支撑的

Notes

[1] With the development of new energy market, China started to phase out NEV subsidies since 2016. It's obvious that the new energy subsidy has been shrinking. Moreover, the government will raise the bar of subsidy approval.

【译文】随着新能源市场的不断发展，从 2016 年开始，我国开始逐步削减新能源补贴。很明显针对新能源汽车的补贴已经缩水，而且政府将提高申请补贴的准入门槛。

【注释】2018 年，财政部、工信部、科技部、国家发展改革委联合发布的《关于调整完善新能源汽车推广应用财政补贴政策的通知》，对计算单车的补贴标准门槛及计算更为复杂：除了要满足续航里程、电池能量密度，还引入了能耗调整系数。按照续航里程的不同，获得不同的补贴金额。续航里程越长，获得的补贴就越高。

[2] A vehicle-to-grid (V2G) system would take advantage of the fact that most vehicles are parked an average of 95 percent of the time. During such idle times the electricity stored in the batteries could be transferred from the PHEV to the power lines and back to the grid.

【译文】V2G 充分利用了大多数汽车平均 95%的时间都处于停泊状态这一事实，在不运行的时间里，可以将储存在插电式混合动力电动汽车电池里的电能传送到输电线，使之最终流向电网。

【注释】Vehicle-to-grid：当混合动力电动汽车或纯电动汽车不运行的时候，可以将电池的能量传送到电网。反过来，当电动汽车的电池需要充满时，电池可以从电网获取电能。其核心思想是利用大量电动汽车的储能源作为电网和可再生能源的缓冲。其实现方法有集中式的 V2G 实现方法、自治式的 V2G 实现方法、基于微网的 V2G 实现方法、基于更换电池组的 V2G 实现方法。

Part B Reading Materials

Plug-in hybrid electric vehicle (PHEV) has several benefits compared to conventional internal combustion engine vehicles. They have lower operating and maintenance costs, and produce little or no local air pollution. They reduce dependence on petroleum and may reduce greenhouse gas emissions from the onboard source of power, depending on the fuel and technology used for electricity generation to charge the batteries. Plug-in hybrids capture most of these benefits when they are operating in all-electric mode. Despite their potential benefits, market penetration of plug-in electric vehicles has been slower than expected as adoption faces several hurdles and limitations. The global market share of the light-duty plug-in vehicle segment achieved 0.86% of total new vehicles sales in 2016, up from 0.62% in 2015 and 0.38% in 2014. However, the stock of plug-in electric vehicles represented just 0.15% of the 1.4 billion motor vehicles on the world's roads by the end of 2016.

Plug-in electric vehicles are more expensive than conventional vehicles, and hybrid electric vehicles due to the additional cost of their lithium-ion battery packs. Other factors discouraging the adoption of electric vehicles are the lack of public and private recharging infrastructure and, in the case of all-electric vehicles, drivers' fear of the batteries running out of energy before reaching their destination due to the limited range of existing electric vehicles. Plug-in hybrids eliminate the problem of range anxiety associated to all-electric vehicles, because the combustion engine works as a backup when the batteries are depleted, giving PHEVs driving range comparable to other vehicles with gasoline tanks. Several national and local governments have established tax credits, subsidies, and other incentives to promote the introduction and adoption in the mass market of plug-in electric vehicles depending on their battery size and all-electric range.

Part C Exercises

1. Talk about your views on new energy vehicles.

Unit Five　Material Forming

Lesson 22　Introduction to Material Forming

教学目的和要求

通过本文的学习，了解制造工艺的分类。本文通过一些实例，简单地介绍几种金属切削加工工艺的方法，如锻造、滚轧、粉末挤压、铸造、车削、电火花加工、电气化学加工、线切割及精密冲裁等。

重点和难点

（1）了解材料成型的分类。

（2）掌握金属切削加工工艺的不同方法的专业词汇和表述。

（3）本文涉及的专业术语较多，对于不同的金属切削加工工艺的英文说明，必须具有相关专业知识才能更好地理解和掌握。

Part A　Text

Material Forming Processes as a System

The term material forming refers to a group of manufacturing methods by which the given shape of a workpiece (a solid body) is converted to another shape without change in the mass or composition of the material of the workpiece.

Classification of manufacturing processes Material forming is used synonymous with deformation or deforming and comprises the methods in group II of the manufacturing process classification shown below. The manufacturing processes are divided into six main groups.

Group I-Primary forming: Original creation of a shape from the molten or gaseous state or from solid particles of undefined shape, that is, creating cohesion between particles of the material.

Group II-Deforming: Converting a given shape of a solid body to another shape without change in mass or material composition, that is, maintaining cohesion.

Group III-Separating: Machining or removal of material, that is, destroying cohesion.

Group IV-Joining: Uniting of individual workpieces to form filling subassemblies, filling and impregnating of workpieces, and so on, that is, increasing cohesion between several workpieces.

Group V-Coating: Application of thin layers to a workpiece, for example, galvanizing, painting, coating with plastic foils, that is, creating cohesion between substrate and coating.

Group VI-Changing the material properties: Deliberately changing the properties of the workpiece in order to achieve optimum characteristics at a particular point in the manufacturing process, [1] These methods include changing the orientation of micro-particles as well as their introduction and removal, such as by diffusion, that is, rearranging, adding, or removing particles[1].

In manufacturing technology, particularly in groups I to IV, we are continually faced with the problem of how to manufacture most economically a particular technical product, with specific tolerance requirements, surface structure, and material properties.

In this section, a short description of the process examples will be given. But assembly and joining processes are not described here.

Forging

Forging can be characterized as: mass conserving, solid state of work material (metal), and mechanical primary basic process-plastic deformation. A wide variety of forging processes are used, and Fig.22.1(a) shows the most common of these: drop forging. The metal is heated to a suitable working temperature and placed in the lower die cavity. The upper die is then lowered so that the metal is forced to fill the cavity. Excess material is squeezed out between the die faces at the periphery as flash, which is removed in a later trimming process[2]. When the term forging is used, it usually means hot forging. Cold forging has several specialized names. The material loss in forging processes is usually quite small. Normally, forged components require some subsequent machining, since the tolerances and surfaces obtainable are not usually satisfactory for a finished product. Forging machines include drop hammers and forging presses with mechanical or hydraulic drives. These machines involve simple translatory motions.

Rolling

Rolling can be characterized as: mass conserving, solid state of material, mechanical primary basic process-plastic deformation. Rolling is extensively used in the manufacturing of plates, sheets, structural beams, and so on. Fig.22.1 (b) shows the rolling of plates or sheets. An ingot is produced in casting and in several stages it is reduced in thickness, usually while hot. Since the width of the work material is kept constant, its length is increased according to the reductions. After the last hot-rolling stage, a final stage is carried out cold to improve surface quality and tolerances and to increase strength. In rolling, the profiles of the rolls are designed to produce the desired geometry.

Powder Compaction

Powder compaction can be characterized as: mass conserving, granular state of material, mechanical basic process-flow and plastic deformation. In this context, only compaction of metal powders is mentioned, but generally compaction of molding sand, ceramic materials, and so on, also belong in this category.

In the compaction of metal powders [Fig.22.1(c)] the die cavity is filled with a measured volume of powder and compacted at pressures typically around 500 N/mm^2. During this pressing phase, the particles are packed together and plastically deformed. Typical densities after compaction are 80% of the density of the solid material. Because of the plastic deformation, the particles are *welded* together, giving sufficient strength to withstand handling. After compaction, the components are heat-treated-sintered-normally at 70%～80% of the melting temperature of the material. The atmosphere for sintering must be controlled to prevent oxidation. The duration of the sintering process varies between 30min and 2h. the strength of the components after sintering can, depending on the material and the process parameters, closely approach the strength of the corresponding solid material.

The die cavity, in the closed position, corresponds to the desired geometry. Compaction machinery includes both mechanical and hydraulic presses. The production rates vary between 6 and 100 components per minute.

Casting

Casting can be characterized as: mass conserving, fluid state of material, mechanical basic process-filling of the die cavity. Casting is one of oldest manufacturing methods and one of the best-known processes. The material is melted and poured into a die cavity corresponding to the desired geometry [Fig.22.1(d)]. The fluid material takes the shape of the die cavity and this geometry is finally stabilized by the solidification of the material.

The stages or steps in a casting process are the making of a suitable mold, the melting of the material, the filling or pouring of the material into the cavity, and the solidification. Depending on the mold material, different properties and dimensional accuracies are obtained. Equipment used in a casting process includes furnaces, mold-making machinery, and casting machines.

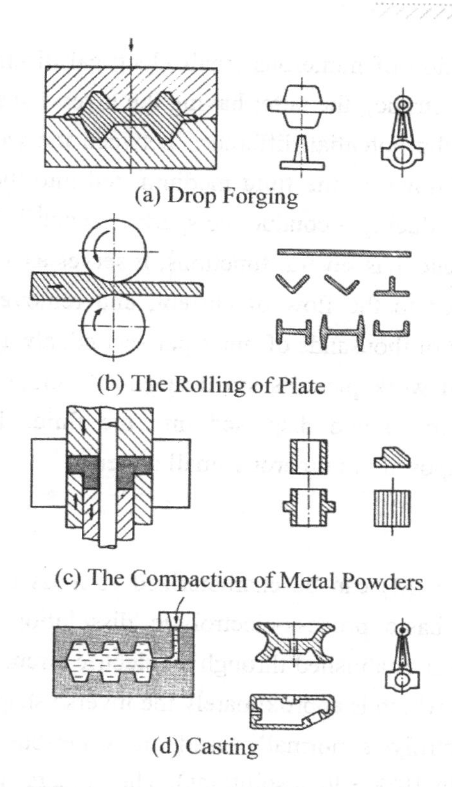

Fig.22.1 Mass-conserving Processes in the Solid State of the Work Material

Turning

Turning can be characterized as mass reducing, solid state of work material, mechanical primary basic process-fracture. The turning process, which is the best known and most widely used mass-reducing process, is employed to manufacture all types of cylindrical shapes by removing material in the form of chips from the work material with a cutting tool [Fig.22.2(a)]. The work material rotates and the cutting tool is fed longitudinally. The cutting tool is much harder and more wear resistant than the work material. A variety of types of lathes are employed, some of which are automatic in operation. The lathes are usually powered by electric motors which, through various gears, supply the necessary torque to the work material and provide the feed motion to the tool.

A wide variety of machining operations or processes based on the same metal-cutting principle are available among the most common are milling and drilling carried out on various machine tools. By varying the tool shape and the pattern of relative work-tool motions, many different shapes can be produced [Fig.22.2 (b) and (c)].

EDM

Electrical discharge machining (EDM) can be characterized as mass reducing, solid state of work material, thermal primary basic process-melting and evaporation [Fig.22.2 (d)]. In EDM, material is removed by the erosive action of numerous small electrical discharges (sparks) between the work material and the tool (electrode), the latter having the inverse shape of the desired geometry. Each discharge occurs when the potential difference between the work material and the tool is large enough to cause a breakdown in the fluid medium, fed into the gap between the tool and work piece under pressure, producing a conductive spark channel[3]. The fluid medium, which is normally mineral oil or kerosene, has several functions. It serves as a dielectric fluid and coolant, maintains a uniform resistance to the flow of current, and removes the eroded material. The sparking, which occurs at rate of thousands of times per second, always occurs at the point where the gap between the tool and work piece is smallest and develops so much heat that a small amount of material is evaporated and dispersed into the fluid. The material surface has a characteristic appearance composed of numerous small craters.

ECM

Electrochemical machining (ECM) can be characterized as mass reducing, solid state of work material, chemical primary basic process-electrolytic dissolution [Fig.22.2 (e)]. Electrolytic dissolution of the work piece is established through an electric circuit, where the work material is made the anode, and the tool, which is approximately the inverse shape of the desired geometry, is made the cathode. The electrolytes normally used are water-based saline solutions (sodium chloride and sodium nitrate in 10%～30% solutions). The voltage, which usually is in the range 5～20V, maintains high current densities, $0.5A/mm^2$～$2A/mm^2$, giving a relatively high removal rate, $0.5cm^3/min \cdot 1000A$～$6cm^3/min \cdot 1000A$, depending on the work material.

Flame Cutting

Flame cutting can be characterized as mass reducing, solid state of work material, chemical primary basic process-combustion [Fig.22.2 (f)]. In flame cutting, the material (a ferrous metal) is heated to a temperature where combustion by the oxygen supply can start. Theoretically, the heat liberated should be sufficient to maintain the reaction once started, but because of heat losses to the atmosphere and the material, a certain amount of heat must be supplied continuously. A torch is designed to provide heat both for starting and maintaining the reaction. Most widely used is the oxyacetylene cutting torch, where heat is created by the combustion of acetylene and oxygen. The oxygen for cutting is normally supplied through a center hole in the tip of the torch.

The flame cutting process can only by used for easily combustible materials. For other materials, cutting processes based on the thermal basic process-melting have been developed (are cutting, are plasma cutting, etc.). This is the reason cutting under both thermal and chemical basic processes.

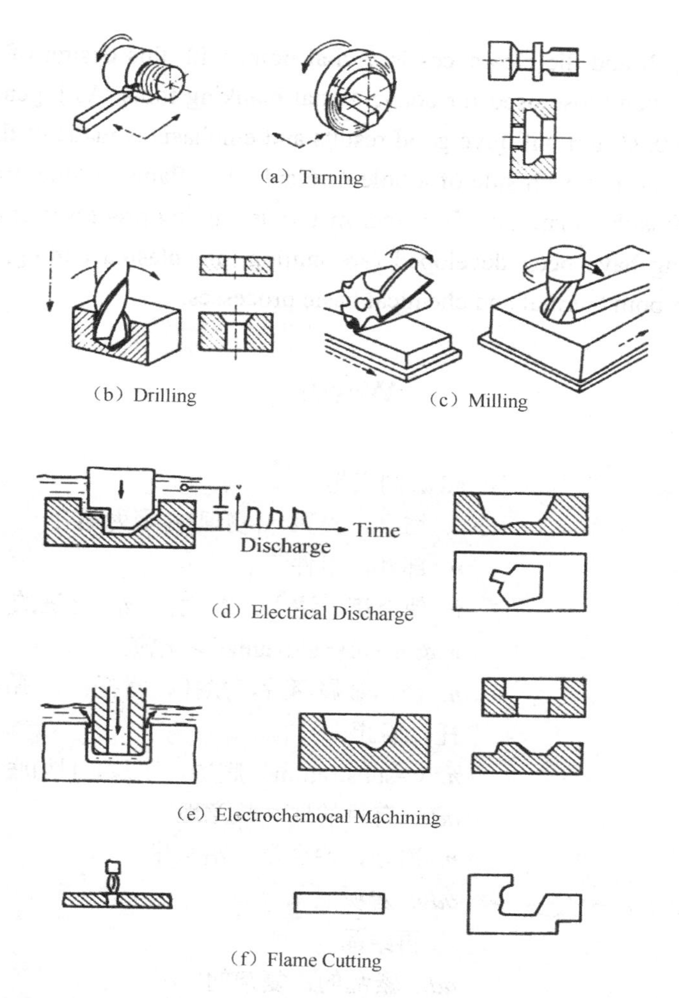

Fig.22.2 Mass-reducing Processes in the Solid State of the Work Material

Fine Blanking

Fine blanking is a technique used for production of blanks perfectly flat and with a cut edge which is comparable to a machined finish. This quick and easy process is worthy of serious thought when the number of parts justifies the cost of a blanking tool especially when consideration is given to the fact that operations such as shaving are eliminated.

One of the fine blanking methods is that in which the punch has a round edge and a small clearance. This is best used for blanks but appears to give less satisfactory results when used for producing holes. In this method the radius on the edge of a die is selected according to the type hardness and thickness of a particular material coupled also to the shape of a profile on the component. The minimum radius that will impart a good result on a component is an essential feature and this radius can vary from 0.3 to 2 mm according to conditions.

The question of punch and die clearances is a vital point with this design of tool and they are always much closer than those used for conventional blanking tools. As a general guide, a total clearance of 0.01 to 0.03 mm will give good results and emphasis is made of the point that these are total clearances and not each side of a hole or blank. The flame cutting process can only by used for easily combustible materials. For other materials, cutting processes based on the thermal basic process-melting have been developed (arc cutting, arc plasma cutting, etc.). This is the reason cutting under both thermal and chemical basic processes.

Words

synonymous	adv. 同义地
cohesion	*n*. 结合，凝聚，[物理]内聚力结合
subassembly	*n*. 部件，组件
impregnate	*v*. 使充满，注入，灌输；*adj*. 充满的
galvanise	*v* .& *n*. <英=galvanize > 电镀
foil	*n*. 箔，金属薄片，烘托，衬托；*v*. 贴金箔于……，衬托，阻止
substrate	*n*.（=substratum）底层，下层，[地]底土层，基础
deliberately	*adv*. 有目的地，故意地
metallurgy	*n*. 冶金，冶金学，冶金术
macroscopically	*adv*. 宏观上
recrystallization	*n*. 再结晶
tangled	*adj*. 紊乱的；复杂的
dislocation	*n*. 位错
polygonized subgrain structure	多边形亚晶结构
grain growth	晶粒长大
squeeze	*v*. 压榨，挤，挤榨
ingot	*n*. [冶]锭铁，工业纯铁
sintering	*v*. 烧结
oxidation	*n*. [化]氧化
solidification	*n*. 凝固
electrolyte	*n*. 电解，电解液
sodium chloride	氯化钠
sodium nitrate	硝酸钠

Notes

[1] These methods include changing the orientation of micro-particles as well as their introduction and removal, such as by diffusion, that is, rearranging, adding, or removing particles.

【注释】

① as well as：和，还有。

② rearranging：重排。

[2] Excess material is squeezed out between the die faces at the periphery as flash, which is removed in a later trimming process.

【注释】

① which：引导定语从句，修饰 excess material。

② as flash：“如闪电一样快速”。

③ trimming process：“清理缝隙的过程”。

[3] Each discharge occurs when the potential difference between the work material and the tool is large enough to cause a breakdown in the fluid medium, fed into the gap between the tool and work piece under pressure, producing a conductive spark channel.

【注释】

① when：引导时间状语从句，修饰 occurs。

② fed into…：从句修饰 fluid medium。

Part B Reading Materials

The so-called ceramic group of cutting tools represents the most recent development in cutting tool materials. They consist mainly of sintered oxides, usually aluminum oxide, and are almost invariably in the form of clamped tips. Because of the comparative cheapness of ceramic tips and the difficulty of grinding them without causing thermal cracking, they are made as throw-away inserts.

Ceramic tools are a post-war introduction and are not yet in general factory use. Their most likely applications are in cutting metal at very high speeds, beyond the limits possible with carbide tools. Ceramics resist the formation of a built-up edge and in consequence produce good surface finishes. Since the present generation of machine tools is designed with only sufficient power to exploit carbide tooling, it is likely that, for the time being, ceramics will be restricted to high-speed finish machining where there is sufficient power available for the light cuts taken. The extreme brittleness of ceramic tools has largely limited their use to continuous cuts, although their use in milling is now possible.

As they are poorer conductors of heat than carbides, temperatures at the rake face are higher than in carbide tools, although the friction force is usually lower. To strengthen the cutting edge, and consequently improve the life of the ceramic tool, a small chamfer or radius is often stoned at the cutting edge, although this increases the power consumption.

Part C Exercises

A brief introduction to the difference between the Flame Cutting and Fine Blanking.

Lesson 23 Introduction to Mould

教学目的和要求

通过本文的学习，了解模具成型的工艺过程，如冲压成型、注塑成型、铸造成型、挤压成型等，以及在不同的成型过程中使用的模具。

重点和难点

（1）了解模具成型的工艺过程。
（2）掌握压成型、注塑成型、铸造成型、挤压成型的专业词汇及其英文表达。
（3）学习本文需要对各种模具成型的工艺过程相关的专业知识和理论有深刻的了解。

Part A Text

Mould is a fundamental technological device for industrial production. Industrially produced goods are formed in moulds which are designed and built specially for them. The mould is the core part of manufacturing process because its cavity gives its shape. There are many kinds of mould, such as casting & forging dies, ceramic moulds, die-casting moulds, drawing dies, injection moulds, glass moulds, magnetic moulds, metal extruding moulds, plastic and rubber moulds, plastic extruding moulds, powder metallurgical moulds, compressing moulds, etc.

The following is the introduction of some of the moulding process and the corresponding moulds used.

Compression Moulding

Compression moulding is the least complex of the moulding processes and is ideal for large parts or low-quantity production. For low-quantity requirements it is more economical to build a compression mould than an injection mould. Compression moulds are often used for prototype, where samples are needed for testing fit and forming into assemblies. This allows for further design modification before building an injection mould for high-quantity production. Compression moulding is best suited for designs where tight tolerances are not required.

A compression mould is simply two plates with cavities cut into either one or both plates. Additional plates between the top and bottom plates can be included to create cavities in the moulded part. Fig.23.1 shown below is an example of a basic two-plate single cavity mold. The mould does not require heater elements or temperature controllers. The moulding temperature is fully controlled by the pressure it is running in.

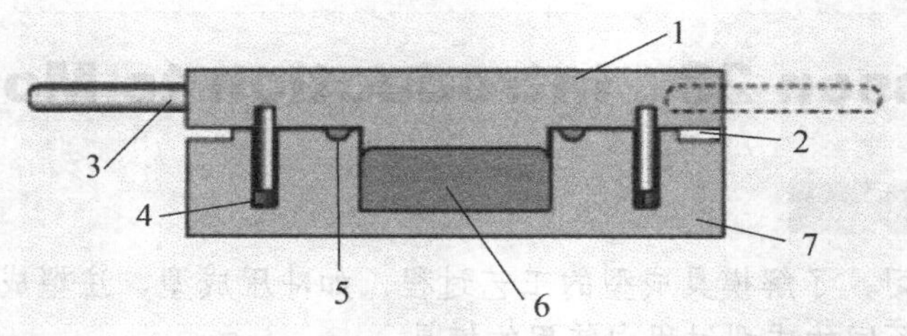

1. top plate 2. opening bar slot 3. handle 4. dowel pin & bushing
flash & tear trim gate 6. part cavity 7. bottom plate

Fig.23.1 A Compression Mould

Due to the simplicity of the mould, it is the most economical mould to buy. And the economical mould/tool cost keeps small quantity running affordable.

Injection Moulding

Injection moulding is the most complex of the moulding processes. Due to the more complex design of the injection mould, it is more expensive to purchase than a cast or compression mould. Although tooling costs can be high, cycle time is much faster than other processes and the part cost can be low, particularly when the process is automated[1]. Injection moulding is well suited for moulding delicately shaped parts because high pressure (as much as 29 000 psi) is maintained on the material to push it into every corner of the mould cavity[2].

Mould is used in injection moulding consist of two halves：one stationary and one movable. The stationary half is fastened directly to the stationary platen and is in direct contact with the nozzle of the injection unit during operation. The movable half of the mould is secured to the movable platen and usually contains the ejector mechanism. The use of a balanced runner system carries the plastic from the sprue to each individual cavity.

An injection mould can be a simpler two plate mould with a runner system to allow the rubber compound to be injected into each cavity from the parting line or a more complex mould with a number of plates, an ejector system and additional heating elements within the core.

Fig.23.2 shown below are examples of basic three-plate and two-plate multi-cavity injection moulds. The moulds do not require heater elements or temperature controllers. The moulding temperature is fully controlled by the injection pressure it is running in.

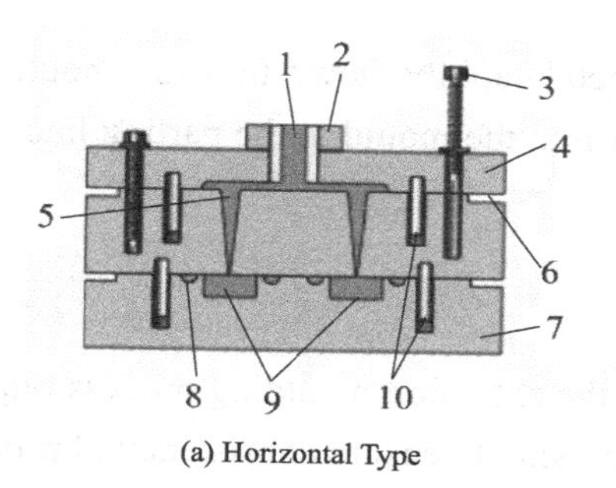

(a) Horizontal Type

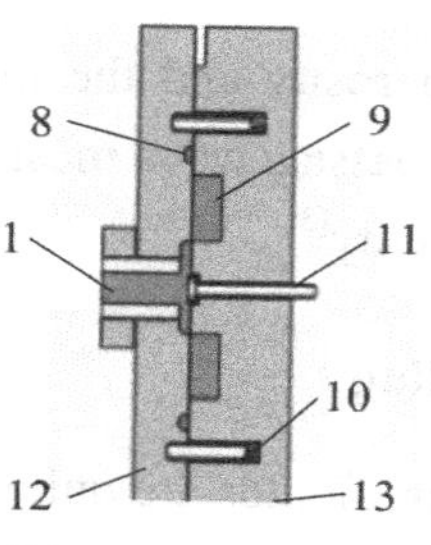

(b) Vertical Type

1. injection runner 2. nozzle bushing 3. stripper bolt 4. top plate 5. sprue
6. opening bar slot 7. bottom plate 8. flash & tear trim gate 9. part cavity
10. dowel pin and bushing 11. ejector 14. fixed plate 13. movable plate.

Fig.23.2 An Injection Mould

Cast Moulding

There are two types of casting, open casting and pressure casting. With open casting, the liquid mixture is poured into the open cavity in the mould and allowed to cure. With pressure casting, the liquid mixture is poured into the open cavity, the cap is put in place and the cavity is pressurized. Pressure casting is used for more complex parts and when moulding foam materials.

In principle, pressure casting is identical to injection moulding with a different class of materials. Cast moulding can, in fact, produce parts that have identical geometries to injection-moulded ones. In many cases, injection moulding has been a substitute for casting moulding due to decreased parts cost. However, for structural parts, particularly those parts with thick-walls, cast moulding can often be the better selection.

Because the materials flow as low-pressure liquids, tooling is generally less expensive. Low tooling costs make casting ideal for small production quantities and prototypes. It is short to medium-production runs. Liquid cast urethane compounds have outstanding resistance to abrasion, impact, and flex fatigue. Also, complex shapes and thick cross-sections can be produced in many compounds. But this process has longer cycle and cure times than other moulding processes does. And once the material has been moulded, it can not be regrind and reused.

The construction of the mould for cast moulding is almost identical to that of mould for injection moulding. It consists of two major sections—the ejector half and the cover half which meet at the parting line. The cavities and cores are usually machined into inserts that are fitted into each of these halves. The cover mould half is secured to stationary platen, while the ejector mould half is fastened to the movable platen. The cavity and machining core must be designed so that the mould halves can be pulled away from the solidified casting. Ejector pins are required to remove the part from the mould when it opens. Lubricants must also be sprayed into the cavities to prevent sticking.

Moulds are usually made of tool steel, mould steel or maraging steel. Since the mould materials have no natural porosity and the molten metal rapidly flows into the mould during injection, venting holes and passageways must be built into the mould at the parting line to evacuate the air and gases in the cavity[3].

Extrusion Moulding

Although extrusion moulds are quite simple the extrusion moulding process requires great care in the setting up and manufacture and final processing to ensure consistency of product[4]. Pressure is forced through the die plate that has the correct profile cut into it. Variations in feed rate, temperature and pressure need to be controlled.

Unlike compression or injection moulding the rubber is not cured when it comes out of the mould. The raw rubber is laid out on circular or long trays (depending on the profile) and loaded into an autoclave for curing under heat and pressure.

For long continuous lengths a salt bath curing system may be used and for silicone extrusion a continuous heating line is used. The curing process used is dependent upon the quantity and profile of the extrusion required.

Most extrusion moulds are simply one round piece of steel with the profile of the intended extrusion wire cut into them. Allowances are made for the shrinkage, expansion of the intended compound. Extrusion dies are the least complex of the moulds.

These moulds are relatively cheap to build but because of the processing involved minimum run quantities will vary.

Words

mould	*n.* 模具，模（型），模塑，压模
cavity	*n.* 模腔，型腔，空洞
die	*n.* 模（子，片，具），压模，塑模
casting & forging die	铸模或锻模
ceramic mould	陶瓷模
die-casting mould	压铸模
drawing die	拉丝模
injection mould	注塑模
magnetic mould	磁铁成型模
extruding mould	挤压成型模
powder metallurgical mould	粉末冶金模
compressing mould	冲压模

moulding	*n.* 成型
prototype	初始制模
assembly	*n.* 装配件，组件
ejector	*n.* 脱模销，推顶器
runner	*n.* 浇道，流道
parting line	合模线，拼缝线
core	*n.* 模芯，中间层
cure	*v.* 固化，塑化
urethane	*n.* 聚氨酯
flex fatigue	弯曲疲劳
maraging steel	马氏体时效钢
venting hole	排气孔，通风孔
evacuate	*v.* 排出，抽空
autoclave	*n.* 高压釜
silicone extrusion	硅橡胶挤压
salt bath curing system	盐浴固化系统

Notes

[1] Although tooling costs can be high, cycle time is much faster than other processes and the part cost can be low, particularly when the process is automated.

【注释】

① tooling costs：模具成本。

② particularly when the process is automated：时间状语从句放在后面，用于进一步说明，译为“尤其当工艺过程为自动操作时”。

[2] Injection moulding is well suited for moulding delicately shaped parts because high pressure (as much as 29 000psi) is maintained on the material to push it into every corner of the mould cavity.

【注释】

① is well suited for：最适合用于……

② delicately shaped parts：外形精致的零件。

③ psi：英制压力单位，磅/平方英寸。

[3] Since the mould materials have no natural porosity and the molten metal rapidly flows into the mould during injection, venting holes and passageways must be built into the moulds at the parting line to evacuate the air and gases in the cavity.

【注释】
① since 引导状语从句，表示原因，译为“由于……”
② venting hole：排气孔。
③ parting line：合模线。

[4] Although extrusion moulds are quite simple the extrusion moulding process requires great care in the setting up and manufacture and final processing to ensure consistency of product.
① in the setting up and manufacture and final processing：在制定、制造和最后的加工过程中。
② to ensure consistency of product：译为“以确保产品设计与制造相一致”，表示目的。

Part B Reading Materials

The most common types of moulds used in industry today are (1) two-plate moulds, (2) three-plate moulds, (3) side-action moulds, and (4) unscrewing moulds.

A two-plate mould consists of two active plates, into which the cavity and core inserts are mounted. In this mould type, the runner system, sprue, runners, and gates solidify with the part being moulded and are ejected as a single connected item. Thus the operation of a two-plate mould usually requires continuous machine attendance.

The three-plate mould consists of: (1) the stationary or runner plate, which contains the sprue and half of the runner. (2) the middle or cavity plate, which contains the other half of the runner, the gates, and cavities and is allowed to float when the mould is open. and (3) the movable or core plate, which contains the cores and the ejector system. This type of mould design facilitates separation of the runner system and the part when the mould opens.

Lesson 24 Mould Design and Manufacturing

教学目的和要求

通过本文的学习，了解 CAD 和 CAM 技术在模具设计和制造中的应用及其优势。

重点和难点

（1）了解 CAD 和 CAM 技术的主要内容。

（2）CAD 和 CAM 技术在模具的设计和制造中所发挥的作用。

（3）理解和掌握 CAD 和 CAM 的相关术语，以及模具设计与制造过程中的相关专业词汇及其英文表达。

Part A Text

CAD and CAM are widely applied in mould design and mould making. CAD allows you to draw a model on screen, then view it from every angle using 3-D animation and, finally, to test it by introducing various parameters into the digital simulation models (pressure, temperature, impact, etc.)[1]. CAM, on the other hand, allows you to control the manufacturing quality. The advantages of these computer technologies are legion: shorter design times (modifications can be made at the speed of the computer), lower cost, faster manufacturing, etc. This new approach also allows shorter production runs, and to make last-minute changes to the mould for a particular part. Finally, also, these new processes can be used to make complex parts.

Computer-Aided Design (CAD) of Mould

Traditionally, the creation of drawings of mould tools has been a time-consuming task that is not part of the creative process. Drawings are an organizational necessity rather than a desired part of the process.

Computer-Aided Design (CAD) means using the computer and peripheral devices to simplify and enhance the design process. CAD systems offer an efficient means of design, and can be used to create inspection programs when used in conjunction with coordinate measuring machines and other inspection equipment. CAD data also can play a critical role in selecting process sequence.

A CAD system consists of three basic components: hardware, software, users. The hardware components of a typical CAD system include a processor, a system display, a keyboard, a digitizer, and a plotter. The software component of a CAD system consists of the programs which allow it to perform design and drafting functions. The user is the tool designer who uses the hardware and software to perform the design process.

Based on the 3D data of the product, the core and cavity have to be designed first. Usually the designer begins with a preliminary part design, which means the work around the core and cavity could change. Modern CAD systems can support this with calculating a split line for a defined draft direction, splitting the part in the core and cavity side and generating the run-off or shut-off surfaces. After the calculation of the optimal draft of the part, the position and direction of the cavity, slides and inserts have to be defined. Then, in the conceptual stage, the positions and the geometry of the mould components such as slides, ejection system, etc. are roughly defined. With this information, the size and thickness of the plates can be defined and the corresponding standard mould can be chosen from the standard catalog. If no standard mould fits these needs, the standard mould that comes nearest to the requirements is chosen and changed accordingly by adjusting the constraints and parameters so that any number of plates with any size can be used in the mould. Detailing the functional components and adding the standard components complete the mould (Fig.24.1). This all happens in 3D. Moreover, the mould system provides functions for the checking, modifying and detailing of the part. Already in this early stage, drawings and bill of materials can be created automatically.

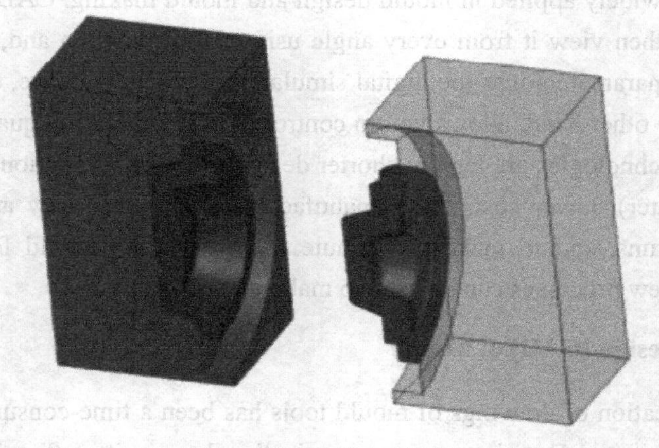

Fig.24.1 3D Solid Model of Mould

Through the use of 3D and the intelligence of the mould design system, typical 2D mistakes such as a collision between cooling and components/cavities or the wrong position of a hole can be eliminated at the beginning. At any stage a bill of materials and drawings can be created allowing the material to be ordered on time and always having an actual document to discuss with the customer or a bid for a mould base manufacturer[2].

The use of a special 3D mould design system can shorten development cycles, improve mould quality, enhance teamwork and free the designer from tedious routine work. The economical success, however, is highly dependent upon the organization of the workflow. The development cycles can be shortened only when organizational and personnel measures are taken. The part design, mould design, electric design and mould manufacturing departments have to consistently work together in a tight relationship.

Computer-Aided Manufacturing (CAM) of Mould

One way to reduce the cost of manufacturing and reduce lead-time is by setting up a manufacturing system that uses equipment and personnel to their fullest potential. The foundation for this type of manufacturing system is the use of CAD data to help in making key process decisions that ultimately improve machining precision and reduce non-productive time. This is called as computer-aided manufacturing (CAM). The objective of CAM is to produce, if possible, sections of a mould without intermediate steps by initiating machining operations from the computer workstation[3].

With a good CAM system, automation does not just occur within individual features. Automation of machining processes also occurs between all of the features that make up a part, resulting in tool-path optimization. As you create features, the CAM system constructs a process plan for you. Operations are ordered based on a system analysis to reduce tool changes and the number of tools used.

On the CAM side, the trend is toward newer technologies and processes such as micro milling to support the manufacturing of high-precision injection moulds with complex 3D structures and high surface qualities. CAM software will continue to add to the depth and breadth of the machining intelligence inherent in the software until the CNC programming process becomes completely automatic. This is especially true for advanced multifunction machine tools that require a more flexible combination of machining operations. CAM software will continue to automate more and more of manufacturing's redundant work that can be handled faster and more accurately by computers, while retaining the control that machinists need.

With the emphasis in the mould making industry today on producing moulds in the most efficient manner while still maintaining quality, mould makers need to keep up with the latest software technologies-packages that will allow them to program and cut complex moulds quickly so that mould production time can be reduced[4]. In a nutshell, the industry is moving toward improving the quality of data exchange between CAD and CAM as well as CAM to the CNC, and CAM software is becoming more *intelligent* as it relates to machining processes resulting in reduction in both cycle time and overall machining time. Five-axis machining also is emerging as a *must-have* on the shop floor-especially when dealing with deep cavities. And with the introduction of electronic data processing (EDP) into the mould making industry, new opportunities have arisen in mould-making to shorten production time, improve cost efficiencies and achieve higher quality.

Words

screen	*n.* 屏幕，隔板
animation	*n.* 动画
digital simulation model	数字模拟模型
legion	*n.* 多，大批，无数
peripheral	*adj.* 外围的，周边的
split	*adj.* 分割的，对分的
run-off	*n.* 流出口，流放口
shut-off	*n.* 截流，断流
ejection system	脱模系统
collision	*n.* 打击，碰撞
tedious	*adj.* 沉闷的
lead-time	*n.* 研制周期
tool-path	*n.* 方法路径
multifunction	*n.* 多功能
redundant	*adj.* 多余的，冗余的
in a nutshell	总之

Notes

[1] CAD allows you to draw a model on screen, then view it from every angle using 3D animation and, finally, to test it by introducing various parameters into the digital simulation models (pressure, temperature, impact, etc.).

【注释】

① draw，view 和 test 表示 3 个并列的操作。

② digital simulation model：数字模拟模型。

[2] At any stage a bill of materials and drawings can be created—allowing the material to be ordered on time and always having an actual document to discuss with the customer or a bid for a mould base manufacturer.

【注释】

① a bill of materials：材料清单。

② allowing 和 having 引导两个现代分词短语，作伴随状语。

[3] The objective of CAM is to produce, if possible, sections of a mould without intermediate steps by initiating machining operations from the computer workstation.

【注释】

① if possible：插入语，译为“如可能”，是 if it is possible 的省略语。试比较以下两个例句：

If possible, I will visit you in Chicago next month.

I will lend you some money to help you with the present difficulty if possible.

② by：通过……方式。

[4] With the emphasis in the mould making industry today on producing moulds in the most efficient manner while still maintaining quality, mould makers need to keep up with the latest software technologies-packages that will allow them to program and cut complex moulds quickly so that mould production time can be reduced.

【注释】

① With：译为“随着……”，引导的介词短语作状语，修饰整个句子。试比较以下例句：

With the development of the economy in China, the people around the country are living a happy life.

② in the most efficient manner 修饰 producing moulds。

③ keep up with：紧跟。

④ that：在这里引导了一个宾语从句，修饰前面的 software technologies-packages。

Part B　Reading Materials

Translate the Following Paragraphs.

A key decision early in mold making process is determining what machining operations will be used and in what order. Machining considerations should be analyzed during the development of the CAD model. If this isn't done, the programmer may not be able to use certain machining strategies.

Each of the processes has advantages and disadvantages when producing a close tolerance mould. The proper selection of process and sequence of process will not only result in more precise dimensional control, but also will reduce manufacturing time by reducing bench work.

The worst case here is that the model may have to be modified, significantly adding to lead-time. Not all workpieces are suitable for hard milling, for example. The smallest internal radius, the largest working depth and the hardness of the material all have to be considered when making this decision.

CAD data can be used to program electrode machining operations. Dimensional data can be downloaded to software that automates electrode design and generates simulations of electrode action so that users can test cut prior to burning. The software also allows users to try different qualities of graphite to determine the optimum grade before actual burning begins.

Part C Exercises

What are the aspects in that its degree of automation is reflected with a good CAM system?

Lesson 25 Heat Treatment of Metal

教学目的和要求

通过本文的学习，了解金属的热处理的定义、相变曲线图的分析、热处理工艺的分类以及不同热处理工艺，如正火、退火、淬火、回火等的概念及特点。

重点和难点

（1）了解金属的热处理的主要内容。
（2）掌握不同的热处理工艺的概念及特点的英文表达。
（3）理解不同热处理工艺之间的差异。

Part A Text

The generally accepted definition for heat treating metals and metal alloys is "heating and cooling a solid metal or alloy in a way so as to obtain specific conditions and I or properties." Heating for the sole purpose of hot working (as in forging operations) is excluded from this definition. Likewise, the types of heat treatment that are sometimes used for products such as glass or plastics are also excluded from coverage by this definition.

Transformation Curves

The basis for heat treatment is the time-temperature-transformation curves or TTT curves where, in a single diagram all the three parameters are plotted. Because of the shape of the curves, they are also sometimes called C-curves or S-curves.

To plot TTT curves, the particular steel is held at a given temperature and the structure is examined at predetermined intervals to record the amount of transformation taken place. It is known that the eutectoid steel (T80) under equilibrium conditions contains, all austenite above 723℃,whereas below, It is pearlite[1]. To form pearlite, the carbon atoms should diffuse to form cementite. The diffusion being a rate process, would require sufficient time for complete transformation of austenite to pearlite. From different samples, it is possible to note the amount of the transformation taking place at any temperature. These points are then plotted on a graph with time and temperature as the axes[2]. Through these points, transformation curves can be plotted as shown in Fig.25.1 for eutectoid steel. The curve at extreme left represents the time required for the transformation of austenite to pearlite to start at any given temperature. Similarly, the curve at extreme right represents the time required for completing the transformation. Between the two curves are the points representing partial transformation. The horizontal lines MS and Mf represent the start and finish of martensitic transformation.

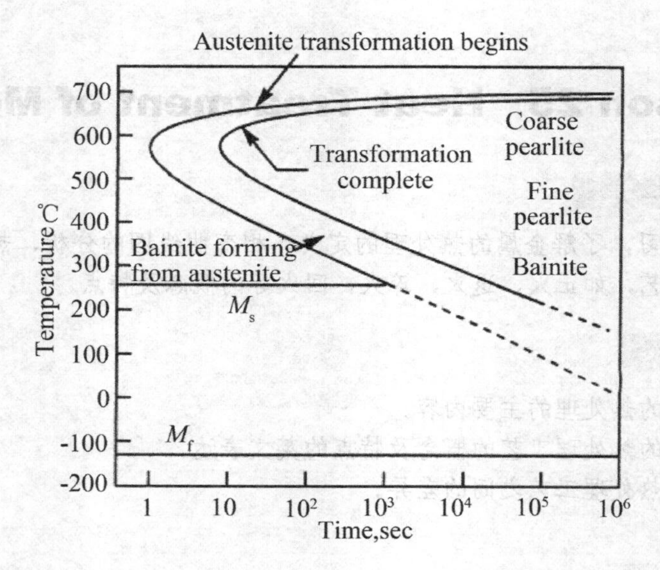

Fig.25.1 Isothermal Decomposition Diagram of T80 Steel

Classification of Heat-Treating Processes

In some instances, heat treatment procedures are clear cut in terms of technique and application. whereas in other instances, descriptions or simple explanations are insufficient because the same technique frequently may be used to obtain different objectives. For example, stress relieving and tempering are often accomplished with the same equipment and by use of identical time and temperature cycles. The objectives, however, are different for the two processes.

The following descriptions of the principal heat treating processes are generally arranged according to their interrelationships.

Normalizing consists of heating a ferrous alloy to a suitable temperature (usually 50°F to 100°F or 28℃ to 56℃) above its specific upper transformation temperature. This is followed by cooling in still air to at least some temperature well below its transformation temperature range[3]. For low-carbon steels, the resulting structure and properties are the same as those achieved by full annealing for most ferrous alloys, normalizing and annealing are not synonymous.

Normalizing usually is used as a conditioning treatment, notably for refining the grains of steels that have been subjected to high temperatures for forging or other hot working operations. The normalizing process usually is succeeded by another heat-treating operation such as austenitizing for hardening, annealing, or tempering.

Annealing is a generic term denoting a heat treatment that consists of heating to and holding at a suitable temperature followed by cooling at a suitable rate. It is used primarily to soften metallic materials, but also to simultaneously produce desired changes in other properties or in microstructure. The purpose of such changes may be, but is not confined to, improvement of machinability, facilitation of cold work (known as in-process annealing), improvement of mechanical or electrical properties, or to increase dimensional stability. When applied solely to relieve stresses, it commonly is called stress-relief annealing, synonymous with stress relieving.

When the term *annealing* is applied to ferrous alloys without qualification, full annealing is implied. This is achieved by heating above the alloy's transformation temperature, then applying a cooling cycle which provides maximum softness. This cycle may vary widely, depending on composition and characteristics of the specific alloy.

Quenching is the rapid cooling of a steel or alloy from the austenitizing temperature by immersing the workpiece in a liquid or gaseous medium. Quenching media commonly used include water, 5% brine, 5% caustic in an aqueous solution, oil, polymer solutions, or gas (usually air or nitrogen).

Selection of a quenching medium depends largely on the hardenability of the material and the mass of the material being treated (principally section thickness).

The cooling capabilities of the above-listed quenching media vary greatly. In selecting a quenching medium, it is best to avoid a solution that has more cooling power than is needed to achieve the results, thus minimizing the possibility of cracking and warp of the parts being treated. Modifications of the term quenching include direct quenching, fog quenching, hot quenching, interrupted quenching selective quenching, spray quenching, and time quenching.

Tempering In heat treating of ferrous alloys, tempering consists of reheating the austenitized and quench-hardened steel or iron to some preselected temperature that is below the lower transformation temperature (generally below 1300°F or 705℃). Tempering offers a means of obtaining various combinations of mechanical properties. Tempering temperatures used for hardened steels are often no higher than 300°F (150℃). The term "tempering" should not be confused with either process annealing or stress relieving. Even though time and temperature cycles for the three processes may be the same, the conditions of the materials being processed and the objectives may be different.

Stress Relieving. Like tempering, stress relieving is always done by heating to some temperature below the lower transformation temperature for steels and irons. For nonferrous metals, the temperature may vary from slightly above room temperature to several hundred degrees, depending on the alloy and the amount of stress relief that is desired.

The primary purpose of stress relieving is to relieve stresses that have been imparted to the workpiece from such processes as forming, rolling, machining or welding. The usual procedure is to heat workpieces to the pre-established temperature long enough to reduce the residual stresses (this is a time-and temperature-dependent operation)to an acceptable level, this is followed by cooling at a relatively slow rate to avoid creation of new stresses.

Words

forge	*v.* 锻造
eutectoid	*adj.* 共析的
austenite	*n.* 奥氏体
pearlite	*n.* 珠光体
martensitic	*adj.* 马氏体的
stress relieving	消除应力，低温退火
tempering	*n.* 回火
normalizing	*n.* 常化，正火
ferrous alloy	铁合金
transformation	*n.* 变换，转换，相变
full annealing	完全退火
notably	*adv.* 显著地，特别是
austenitize	*v.* 奥氏体化，使成奥氏体
machinability	*n.* 切削加工性，机械加工性能
facilitation	*n.* 便于
quenching	*n.* 淬火
brine	*n.* 盐水
caustic	*adj.* 腐蚀性的，碱性的
aqueous	*adj.* 水的，水成的
warp	*n.* 翘曲，变形
glossary	*n.* 词汇表，术语汇编
quench-hardened	*adj.* 淬火硬化的
process annealing	工序间退火，中间退火
fog quenching	喷雾淬火
interrupted quenching	分级淬火
selective quenching	局部淬火

Notes

[1] It is known that the eutectoid steel (C80) under equilibrium conditions contains, all austenite

above 723℃, whereas below, it is pearlite.
【译文】我们知道共析钢（C80）在平衡条件下，在723℃以上时全为奥氏体，而低于此温度，则为珠光体。
【注释】
① that：引导主语从句，译为“我们知道”。
② eutectoid steel：共析钢。
③ below：指低于723℃。

[2] These points are then plotted on a graph with time and temperature as the axes.
【注释】
① with time and temperature as the axes：axes，轴，此处为坐标轴，以时间和温度为坐标轴。

[3] This is followed by cooling in still air to at least some temperature well below its transformation temperature range.
【注释】
① follow：接着，跟着。
② 句中的this代替上一句话的内容，表示先“heating a ferrous alloy…”，再“cooling…”。

Part B Reading Materials

Translate the Following Paragraphs.

Carburizing consists of absorption and diffusion of cartoon solid ferrous alloys by heating to some temperature above the upper transformation temperature of the specific alloy. Temperatures used for carburizing are generally in the range of 1650° F of to 1900°F (900℃ to 1040℃). Heating is done in a carbonaceous environment (liquid, solid, or gas). This produces a carbon gradient extending inward from the surface, enabling the surface layers to be hardened to a high degree either by quenching from the carburizing temperature or by cooling to room temperature followed by reaustenitizing and quenching.

Carbonitriding is a case hardening process in which a ferrous material (most often a low-carbon grade of steel) is heated above the transformation temperature in a gaseous atmosphere of such composition as to cause simultaneous absorption of carbon and nitrogen at (by) the surface and, by diffusion a concentration, gradient is created. The process is completed by cooling at a rate that produces the desired properties in the workplace.

Part C Exercises

What is the primary purpose of stress relieving?

Unit Six Advanced Manufacturing Technology

Lesson 26 Additive Manufacturing

教学目的和要求

本文为介绍增材制造的英文专业文献，主要围绕增材制造及其工艺技术的定义与专业英文表达进行阐述。通过本文的学习，读者可以了解有关增材制造中常见专业词汇的英文表达和专业术语的英文名称，有助于读者全面了解增材制造过程及其所涉及的专业知识；要求掌握文中所涉及的增材制造相关术语、相关学科的英文表达，并能对文后所附的练习进行理解和翻译。

重点和难点

（1）重点掌握增材制造相关的专业术语及其英文表达。

（2）了解增材制造的相关定义及其工艺技术方法。

Part A Text

The Additive Manufacturing

3D printing technology was born in the United States in the 1980s. China began researching 3D printing technology in 1991. The name was called rapid prototyping technology, which was the physical model before the development of samples[1]. In order to facilitate the promotion of rapid prototyping system technology and the acceptance of the public, the industry has adopted this kind of scientific and technical system based on the principle of discrete-stacking, which directly drives parts by three-dimensional data of parts, collectively referred to as three-dimensional printing, that is, three-dimensional printing in a broad sense[2]. However, it is called additive manufacturing in academic and government documents at home and abroad.

Additive manufacturing is relative to the reduction manufacturing used in traditional manufacturing. Reduced material manufacturing is to shape, cut, and remove raw materials through machining methods such as molds, turning and milling, and finally produce finished products. Contrary to the manufacturing method of the reduced material, the additive manufacturing is a technique of manufacturing solid parts by gradually adding materials. It transforms the three-dimensional entity into several two-dimensional planes, and by mass-processing and layer-by-layer stacking, it is using Brick wall, adding materials layer by layer, and finally forming objects. It is a "bottom-up" manufacturing method that greatly reduces the complexity of manufacturing.

Additive manufacturing technology has five major technical features: digital manufacturing, dimensionality reduction manufacturing, stacked manufacturing, direct manufacturing, and rapid manufacturing. The core of additive manufacturing technology is the combination of digital, intelligent manufacturing and materials science. It is based on computer three-dimensional design model, through software layered discrete and numerical control molding system, using laser beam, hot melt nozzle, etc. Special materials such as ceramic powder, plastic, and cell tissue are layer-by-layer stacked and bonded to form a solid product.

Process technology

The main processes of additive manufacturing technology are: stereo lithography, layered solid fabrication, selective laser sintering, fused deposition molding, laser cladding, and patternless molding.

Stereo lithography

Stereo lithography, also called light curing, light modeling. This process was invented by Charles Hull in the United States in the 1980s. In 1986, 3D Systems of the United States introduced the commercial prototype SLA-1, the world's first rapid prototyping system. The process is filled with liquid photosensitive resin as a material, and the computer controlled laser beam tracks the layered cross-section track and irradiates the liquid resin in the liquid tank to solidify the layer of resin, and then the lifting platform descends to a height. The formed layer is covered with a layer of resin, and then a new layer is scanned. The newly cured layer is firmly adhered to the previous layer, and is repeated until the entire part is manufactured to obtain a three-dimensional solid model. The characteristics of the process are high precision of the prototype parts, good strength and hardness of the parts, and the hollow parts with complex shapes can be produced. The model of the production is flexible and can be disassembled at will, which is an ideal method for indirect molding[3]. The disadvantage is that support is required, the shrinkage of the resin leads to a decrease in precision, and the resin has a certain toxicity and does not conform to the development trend of green manufacturing.

Layered entity manufacturing

Layered entities are manufactured, and data is also translated into laminated entities. This process was successfully developed in 1986 by Michael Feijin Helisys, USA. The principle of the process is to cut the foil and paper according to the layered geometric information of the part, and bond obtained layer into a three-dimensional entity. The process is first to lay a layer of foil, such as paper, plastic film, etc., and then use the laser to cut the outline of the layer under computer control, and the non-part parts are all chopped for easy removal. When the layer is completed, a layer of foil is placed, rolled and heated with a roller to cure the binder, so that the newly laid layer is firmly bonded to the formed body, and then the contour of the layer is cut. This is repeated until the processing is completed, and finally the chopped portion is removed to obtain a complete part. The process is characterized by reliable operation, good model support, low cost and high efficiency. The disadvantage is that the front and back processing is time consuming and laborious, and the hollow structural members cannot be manufactured. Since the process material is limited to paper or plastic film, the performance has not improved, and it has gradually entered the decline. Most organizations have already or are about to give up the process.

Selective laser sintering

Selective laser sintering, as well as data translation into laser selective sintering. This process was first developed in 1989 by Dekad, a researcher at the University of Texas at Austin. The materials commonly used in this process are powders of metals, ceramics, ABS plastics, etc. as molding materials. The process is to first lay a layer of powder on the workbench, selectively sinter with a laser beam under computer control, and the sintered portion is solidified to form a solid part of the part. After the first layer is completed, the next layer is carried out, and the new layer is firmly sintered with the upper layer. After all the sintering is completed, the excess powder is removed to obtain a sintered part. The process is characterized by a wide range of materials, not only for the manufacture of plastic parts, but also for the manufacture of ceramic, metal, wax and other materials.

Fused deposition

Fused deposition molding, also translated into mixed deposition modeling, melt extrusion molding. This process was first proposed by American scholars in 1988, and the United States Stratasys developed the first commercial model in 1992. The process is based on the thermoplastic molding material wire, which is melted into a liquid through the extrusion head of the heater, and the computer controls the extrusion head to accurately move along the contour of each section of the part, so that the molten thermoplastic material wire passes. The nozzle is extruded, overlaid on the finished part, and solidified rapidly in a very short time to form a layer of material. Thereafter, the extrusion head is moved upward in the axial direction by a slight distance to construct the next layer of material. This stacks into a solid model or part from bottom to top. The process is characterized by simple use and maintenance, low manufacturing cost and high speed. The general complexity prototype can be molded in a few hours without pollution.

Laser cladding process

The laser cladding process utilizes a high-energy density laser beam to rapidly melt the surface of the alloy having different compositions and properties and the surface of the substrate to form a rapid solidification process on the surface of the substrate to form an alloy layer having completely different compositions and properties from the substrate.

Patternless mold manufacturing

The patternless mold manufacturing was developed in 1997 by the Laser Rapid Prototyping Center of Tsinghua University. The process is to first obtain a cast CAD model from the part computer aided design model. The cross-sectional profile information is obtained by computer layering, and the control information is generated by the layer information. In the modeling, the first nozzle sprays the binder on each layer of the paved sand, and the second nozzle sprays the catalyst along the same path. The two colloidal reactions occur, and the layers are solidified and stacked. The metal can be cast after the inner surface of the resulting sand mold is coated or impregnated with the coating. The process is characterized by short manufacturing time, no need for wooden molds, integrated molding, simultaneous molding of the core and core, and the production of the molds with curved surfaces and curves.

Words

manufacturing	*n.* 制造业，工业；*adj.* 制造业的，制造的
prototyping	*n.* 原型法；原型；样机；成型
physical	*adj.* 自然（界）的；身体的；物质的
abroad	*adv.* 到国外，在海外；广为流传地
relative	*adj.* 相关的；相对的；相互有关的
milling	*n.* 铣床；研磨；铣削刀具；铣
materials	*n.* 材料；[哲学] 物质
software	*n.* 软件；软体
track	*n.* 小路；踪迹；方针，路线；*vt.* 跟踪；监看，监测；追踪
previous	*adj.* 先前的；以前的；过早的
flexible	*adj.* 灵活的；易弯曲的；柔韧的
obtained	*v.* 获得
surfaces	*n.* 曲面，表面
impregnated	*v.* 灌注，使饱和
deposition	*n.* 沉积（物）
ceramic	*n.* 烤瓷
laser	*n.* 激光；激光器，镭射器

Notes

[1] The name was called rapid prototyping technology, which was the physical model before the development of samples.

【注释】此句为长句子，其中，which 引导宾语从句，指代“rapid prototyping technology”。

[2] In order to facilitate the promotion of rapid prototyping system technology and the acceptance of the public, the industry has adopted this kind of scientific and technical system based on the principle of discrete-stacking, which directly drives parts by three-dimensional data of parts, collectively referred to as three-dimensional printing, that is, three-dimensional printing in a broad sense.
【注释】此句为长句子，其中，which 引导宾语从句，指代“the principle of discrete-stacking”。

[3] The model of the production is flexible and can be disassembled at will, which is an ideal method for indirect molding.
【注释】此句是省略句，其中，which 指代主语“The stereo lithography”。

Part B Reading Materials

Additive Manufacturing a tool for innovation and entrepreneurship

Additive manufacturing (3D printing) technology is considered “a technology that will change the world.” The British “Economist” magazine believes that additive manufacturing will “cooperate with other digital production models to promote the third industrial revolution.” In 2013, McKinsey released *Vision 2025*, and additive manufacturing was included in one of the 12 major disruptive technologies that determine the future economy. Additive manufacturing technology provides a historic opportunity for the development and upgrading of China’s manufacturing industry. Additive manufacturing can quickly and efficiently achieve the physical prototype of new products, providing a fast-technical approach to product development. This technology reduces the technical and personnel technology threshold of the manufacturing industry, helps to promote the small and micro manufacturing service industry, effectively raises the level of employment, helps activate social wisdom and financial resources, realizes the structural adjustment of the manufacturing industry, and promotes the change of manufacturing industry.

(1) Open up a huge space for innovation and entrepreneurship. 3D printing is suitable for complex shape structures, multi-variety, small batch manufacturing and applications in many fields. People can maximize the function of materials through topology optimization design and multi-material manufacturing functional gradient structure, which brings disruptive progress to many equipment design and manufacturing, frees the design from the constraints of traditional technology manufacturability, and releases the innovative design.

(2) The new production organization model provides unlimited business opportunities for entrepreneurship. Additive manufacturing brings a new model of distributed manufacturing, that is, through the network platform, individualized orders, maker design, manufacturing equipment, and even integrated planning and decentralized implementation of funds, this production model can effectively achieve the maximum play of social resources. Provide technical support for entrepreneurship and ubiquitous manufacturing.

(3) Promote the revolutionary development of cross-disciplinary research. Through cell printing, tissue engineering, organ reconstruction, through the construction of stem cell test bench, rapid and efficient stem cell induction test, the development of genetic printing, providing a leap forward development for the development of life sciences.

(4) Bringing great opportunities for the development and upgrading of China's manufacturing industry. 3D printing is a tool for product innovation and has become an advanced development model. The overcapacity and the serious lack of product development capacity are the bottlenecks in the development of China's manufacturing industry. The rapid application of 3D printing in various fields is an important means of developing a high-tech service industry, realizing the restructuring of the manufacturing industry and promoting the manufacturing industry from large to strong.

Part C Exercises

Translate the following paragraphs into Chinese.

4D printing technology and its application research in intelligent material structure are still in their infancy. However, its research and development applications will have a profound impact on the manufacture of traditional mechanical structures. Design and this development trend are reflected in the following aspects.

(1) 4D printing smart materials will change the mode of "mechanical transmission + motor drive" in the past. The current mechanical structure system is mainly the transmission mode of mechanical transmission and drive. In the future, the in-situ drive mode of functional materials is no longer restricted by the freedom of motion of the mechanical structure, and can realize continuous freedom and controllable functions of stiffness, while at the same time The weight will also be greatly reduced.

(2) 4D printing technology manufactures a smart material structure that integrates driving and sensing to realize the fusion of driving and sensing performance of smart materials. EAP materials have good driving performance and sensing performance, that is, deformation can occur under the action of electric field, and voltage and current signals can be output as they are deformed. The research team combines the driving performance of EAP materials with sensing performance, and uses 4D printing technology to manufacture a flexible operating arm for driving and sensing integrated minimally invasive surgery. The operating arm can not only bend the operating arm by electric field to drive the deformation of the operating arm, but also In the process of bending deformation, the sensing arm of the intelligent material can be used to accurately and appropriately deform the operating arm without damaging the human tissue, thereby solving the problem that the traditional minimally invasive surgical instrument causes damage to the human

body during the deformation process due to lack of sensing function.

(3) Research and development of a variety of smart materials suitable for 4D printing technology, responding to different external environmental incentives, and responding to more diverse forms of deformation. At present, the incentive modes and deformation forms of 4D printed smart materials are limited. Skylar Tibbits and others are currently researching and developing smart material 4D printing technology that can respond to vibration and sound waves. With the diversification of 4D printing smart materials, the application of 4D printing technology will be more extensive.

Lesson 27 The Robot Technology

教学目的和要求

本文是介绍机器人技术的英文专业文献，主要围绕机器人技术及其应用领域相关技术的定义与专业英文表达进行介绍。通过本文的学习，读者可以了解有关机器人技术中常见专业词汇的英文表达和专业术语的英文名称，有助于全面了解机器人技术所涉及的专业知识；要求掌握文中所涉及的机器人相关术语、相关学科的英文表达，并能对文后所附的练习进行理解和翻译。

重点和难点

（1）重点掌握机器人技术相关的专业术语及其英文表达。
（2）了解机器人的相关定义及其应用领域。

Part A Text

Robot Technology

Since the 21st century, more and more attention has been paid to the development of robotics at home and abroad. Robotics is considered to be one of the high technologies that are important for the development of emerging industries in the future[1]. The domestic and international industry has high hopes for robot technology to lead the development of the future industry. Robot technology is one of the foundations for the development of high-tech and emerging industries in the future, and is of great significance to the national economy and national defense construction.

Driven by the development of new technologies such as computer technology, network technology, and MEMS technology, robotics is rapidly expanding from traditional industrial manufacturing to medical services, education and entertainment, exploration and survey, bioengineering, disaster relief, and agricultural engineering. Robot systems in different fields are being researched and developed.

Industrial Robot

The term “industrial robot” was proposed by the American Metal Market in 1960 and defined by the American Robot Society as “a programmable multi-function operator for handling mechanical parts or work pieces, or by changing programs.” This definition has now been adopted by the International Organization for Standardization.

In general, industrial robots consist of three major parts and six subsystems. 3 is mostly a mechanical part, a sensing part and a control part. The six subsystems can be divided into mechanical structure system, drive system, sensing system, robot- environment interaction system, human-computer interaction system and control system.

From the perspective of mechanical structure, industrial robots are generally divided into series robots and parallel robots[2]. The characteristic of a tandem robot is that the motion of one axis changes the coordinate origin of the other axis[3], and the parallel mechanism used by the parallel robot does not change the coordinate origin of the other axis. Early industrial robots used a series mechanism. The parallel mechanism is defined as a closed loop mechanism in which the moving platform and the fixed platform are connected by at least two independent kinematic chains, the mechanism has two or more degrees of freedom, and is driven in parallel. Compared with the series robot, the parallel robot has the advantages of high rigidity, stable structure, large carrying capacity, high fine motion precision and small motion load. In the position solving, the positive solution of the tandem robot is easy, but the reverse solution is very difficult while the parallel robot is the opposite, the positive solution is difficult, and the reverse solution is very easy.

The driving methods of industrial robots mainly include hydraulic drive, pneumatic drive and motor drive. Motor drive is a mainstream drive method for modern industrial robots. It is divided into 4 types of motors: DC servo motor, AC servo motor, stepper motor and linear motor. DC servo motor and AC servo motor adopt closed-loop control, which is generally used for high-precision and high-speed robot drive; stepper motor is used for occasions where accuracy and speed are not high, and open-loop control is adopted linear motor and its drive control system are that the technology has become increasingly mature, and has the superior performance that traditional transmissions cannot match.

Medical Robot

As early as 1985, researchers completed the robotic assisted positioning of neurosurgical biopsy with PUMA 560 industrial robots. This is the first time that robotics has been used in medical surgery, marking the beginning of the development of medical robots. After decades of rapid development, medical robots have been widely used in neurosurgery, abdominal surgery, thoracic surgery, orthopedic surgery, vascular intervention, craniofacial surgery and other operations. According to Reportlinker, the global market for medical robots in 2013 was $2.7 billion, in 2014 it was $3.3 billion, and in 2019 it was $4.6 billion. From 2014 to 2019, the compound annual growth rate will reach 7%. In the same period, the Asia-Pacific region’s five-year annual growth rate was 13.4%. Therefore, the current development of medical robots faces enormous market opportunities.

Medical machines are classified into neurosurgical robots, orthopedic robots, laparoscopic robots, vascular interventional robots, prosthetic and exoskeleton robots, auxiliary rehabilitation robots, hospital service robots, and capsule robots according to the functions and uses of medical robots. After nearly 30 years of development, robot technology has been successfully applied in many medical fields such as surgery, rehabilitation, hospital services, disease diagnosis, etc. The development of medical robots will continue to change the traditional medical methods.

From the current main research hotspots of medical robots, in order to meet the adaptability and reliability of medical robots in complex environments, medical robots will develop in the direction of intelligence and autonomy, in order to lower the threshold for medical robot development and accelerate its development, there is a need for a unified open source software and hardware platform for medical robots, in order to further reduce surgical trauma and shorten the recovery time of surgery, single-hole and natural-hole surgical robots will become the main solution, in order to further expand the application scope of medical robots, micro-robots will bring a new medical revolution. At the same time, with the increase of the global aging population, the demand for medical robots is also increasing, which also provides a broader market space and development opportunities for the development of medical robots[4].

Agricultural Robot

Agriculture is the foundation of the national economy, and modern agricultural equipment is an important support for modern agriculture. In recent years, with the development of facility agriculture, precision agriculture and high-tech, especially the land transfer and the scale of agricultural production, the intensification, and the rising cost of manual operations, agricultural robots have become a substitute for heavy manual labor and improved production. Conditions, improving production efficiency, transforming development methods, reducing production costs and losses, and enhancing key equipment for comprehensive production capacity are also one of the focus of technological competition in the international agricultural equipment industry.

The agricultural robot is a flexible automated or semi-automated device that uses agricultural products as the object of operation, has both human information perception and limb action functions, and can be reprogrammed. It can reduce labor intensity, solve labor shortage, improve labor productivity and work quality, and prevent pesticides, fertilizers and other damage to the human body. At present, agricultural robots have been greatly developed. Robots can replace artificial agricultural activities, such as spraying pesticides in fields and greenhouses, harvesting and sorting of some crops, and some difficult work for humans, such as high-altitude picking. The widespread use of agricultural robots will greatly change the mode of labor of traditional agriculture, reduce the dependence on a large number of labor, and contribute to the transformation of agriculture from tradition to modernity.

With the emphasis on research and development of agricultural robots at home and abroad in recent years, a variety of agricultural robots have been developed. According to the different focus of solving the problem, agricultural robots can be roughly divided into two categories: one is the walking series of agricultural robots, which are mainly used for working in large-area farmland, the other type of robots are mainly used in greenhouses or plants, working in the workshop.

Intelligent Mobile Robot

Mobile robot is a comprehensive system integrating environment perception, dynamic decision-making and planning, behavior control and execution. It focuses on multi-disciplinary research results in sensor technology, mechanical engineering, electronic engineering, computer engineering, automation control engineering and artificial intelligence. It represents the highest achievement of mechatronics and is one of the most active areas of scientific and technological development. With the continuous improvement of robot performance, the application range of mobile robots has been greatly expanded, not only in industrial, agricultural, defense, medical, service and other industries, but also in the fields of mine clearance, search, rescue, radiation and space. It is very good application with dangerous occasions. Therefore, mobile robot technology has received widespread attention from all over the world.

Intelligent mobile robots have a perceived work environment, mission planning capabilities and decision control capabilities. It includes an architecture of three basic elements of perception, planning, and execution. At present, mobile robots have three kinds of architectures: hierarchical structure, reaction structure and deliberate/reactive hybrid structure. Most of the early mobile robots used a hierarchical structure. The advantage of this architecture is that it gives the order of the relationship between perception, planning and execution, the disadvantage is that the planning part, the perception, the planning algorithm is slow, and the execution response time is long.

In short, mobile robots involve multi-disciplinary technologies such as sensor technology, electronic technology, computer technology, control technology, and artificial intelligence technology. Despite the encouraging results of mobile robots, there is still a long way to go to implement intelligent, emotional mobile robots. With the continuous improvement of electronic technology, computer technology and artificial intelligence technology, we have reason to believe that in the near future, mobile robots have more intelligence and emotions to better serve human beings.

Words

robot	*n.* 机器人；遥控装置
emerging	*adj.* （用作定语）新兴的

industry	*n.*工业、行业
significance	*n.* 意义；意思；重要性
expanding	*adj.* 展开的，扩大的； *v.* 使……变大（expand的现在分词）；扩大；伸展
entertainment	*n.* 娱乐，消遣；招待，款待；娱乐节目
survey	*vt.* 调查；勘测；俯瞰；*n.* 调查（表），调查所，测量
divide	*vt.* 分开；划分
environment	*n.* 环境，外界；周围，围绕；工作平台
parallel	*n.* 并联；平行埠；平行度；平行线
kinematic	*adj.* 运动学的，运动学上的
rigidity	*n.* 坚硬；严格；刚直；死板
tandem	*n.* 串联
hydraulic	*adj.* 水力的，水压的
craniofacial	*adj.* 颅面的
enormous	*adj.* 巨大的；庞大的；极恶的；凶暴的
neurosurgical	*adj.* 神经外科的
capsule	*n.* 胶囊；航天舱；*adj.* 压缩的；概要的；*vt.* 压缩；简述
agricultural	*adj.* 农业的，耕种的；农艺的，农学的
modern	*adj.* 现代的；近代的；新式的；*n.* 现代人，现代主义者

Notes

[1] Robotics is considered to be one of the high technologies that are important for the development of emerging industries in the future.

【注释】此句为 that 从句，句中 that 指代“Robotics”。

[2] From the perspective of mechanical structure, industrial robots are generally divided into series robots and parallel robots.

【注释】此句为 From 引导的从句，为了使句式结构明了，From 从句也是指代主语“industrial robots”的属性。

[3] The characteristic of a tandem robot is that the motion of one axis changes the coordinate origin of the other axis, and the parallel mechanism used by the parallel robot does not change the coordinate origin of the other axis.

【注释】此句为 that 从句，that 从句作为“The characteristic of a tandem robot”的宾语。

[4] …, the demand for medical robots is also increasing, which also provides a broader market space and development opportunities for the development of medical robots.

【注释】此句为 which 从句，which 指代“the demand for medical robots”。

Part B　Reading Materials

Research Status and Development Trend of Service Robot Technology

As a strategic high technology, the service robot technology has a long and strong industrial chain and is still in a stage of decentralized development worldwide. Through the research of core technologies and products of service robots, it is of great significance to the country's major needs and security, through the development of cutting-edge technologies, core components and related standards, it plays an important role in promoting the development of national people's livelihood science and strategic emerging industries. Explorations such as decision-making and execution have an important role in promoting traditional industrial upgrading and services.

On the one hand, as China gradually enters an aging society, the demand for China's old-age service robots will face a spurt-type growth, the old-age service robot products have broad market prospects and huge development potential, providing China with a rare opportunity for the development of the service robot industry. On the other hand, public security events such as earthquakes, floods and extreme weather, as well as mine disasters, fires, social security, etc. in addition, medical and educational demand for service robots is strong. These indicate that China's service robots have a huge potential market to develop.

The service robot technology has the characteristics of comprehensiveness and permeability. It focuses on the use of robot technology to complete the service work beneficial to human beings. It has broad application prospects in the elderly/disabled people and special fields, and has the characteristics of strong radiation and obvious economic benefits. Service robot technology is not only a technical contest for the country's future space, underwater and underground resource exploration, weaponry and commanding heights, but also a strategic emerging industry with high-tech competition between countries, including helping the elderly, helping the disabled, dangerous operations, education and entertainment, etc. It is an important part of the future advanced manufacturing industry and modern service industry, and also a major opportunity for the development of the world's high-tech industry.

Part C　Exercises

Translate the following paragraphs into Chinese

Key Technology Analysis of BigDog Quadruped Robot

The BigDog quadruped robot can be summarized as follows. Mainly based on the quadruped mammal structure as a biomimetic reference, designed and manufactured by pure mechanical method, with 12 or 16 active degrees of freedom of the leg movement device, hydraulically driven system for the active degree of freedom to implement power output, airborne motion control The system can detect the attitude and the terrain of the body, use the virtual model to measure the key parameters such as the position of the weight of the machine, and then use the virtual model to implement correct and safe motion planning, and implement accurate planning and output according to the actual load dynamics of the limb. And the output is adjusted synchronously according to the change of the body state, so that the robot has strong adaptability to complex terrain. BigDog has a high degree of motion autonomy, as well as high navigation intelligence, independent perception of the environment and autonomous planning path, and rarely requires manual intervention. BigDog is a typical unstructured environment four-legged mobile robot with full autonomous motion capability and strong autonomous navigation capability. It is a land mobile robot that is difficult to implement in the current robot field.

Lesson 28 Intelligent Manufacturing

教学目的和要求

本文为介绍智能制造的英文专业文献，主要围绕智能制造的内涵及其应用领域相关技术的定义与专业术语的英文表达进行介绍。通过本文的学习，读者可以了解有关智能制造中常见专业词汇的英文表达和专业术语的英文名称，有助于全面了解智能制造所涉及的专业知识；要求掌握文中所涉及的智能制造相关术语、相关学科的英文表达，并能对文后所附的练习进行理解和翻译。

重点和难点

（1）重点掌握智能制造相关专业术语及其英文表达。
（2）了解智能制造的应用领域及其发展趋势。

Part A Text

Origin of Concept

From the controversy over the nature of the third industrial revolution, more research has taken "manufacturing intelligentization" as the core of the third industrial revolution. It can be considered that intelligent manufacturing will be the leading industry leading the third industrial revolution[1].

Intelligent manufacturing is not a new thing, it is the continuous popularization of information technology and gradually developed in the 80s of 20 century. Advanced manufacturing technology and computer technology are widely used in modern manufacturing[2]. The traditional design methods and management methods cannot effectively solve the new problems in modern manufacturing system. Through the organic integration of traditional manufacturing technology, artificial intelligence science, computer technology and science, we developed a new manufacturing technology and system, that is, intelligent manufacturing technology and intelligent manufacturing system.1988, Professor White of York University and Professor Boone of Carnegie Mellon University published the book of "intelligent manufacturing" for the first time. The idea of intelligent manufacturing is that the purpose of intelligent manufacturing is to model the skills and expert knowledge of the manufacturing technicians by integrating knowledge engineering, manufacturing software systems, robot vision and machine control, so that the intelligent robot can be produced in a small mass without manual intervention.

The connotation of intelligent manufacturing

Intelligent manufacturing not only adopts new manufacturing technology and equipment, but also infiltrates the rapidly developing information and communication technology[3] (Internet of things and service Internet) into the factory, constructs the information physical system in the manufacturing field, thus completely changes the manufacturing organization mode and man-machine relationship of manufacturing industry, and brings about the change of business model. The technical breakthrough in a certain field is not simply using information technology to transform traditional industries, but the integration and integration of information and communication technology (ICT) and manufacturing.

Intelligent manufacturing is a new type of manufacturing system, which combines advanced automation technology, sensing technology, control technology[4], digital manufacturing technology and Internet of things, large data, cloud computing and other new generation of information technology to realize the real time management and optimization of the whole life cycle of the factory and enterprise, and the whole life cycle of the product, which can reduce the production cost to the maximum limit. To reduce energy resources consumption, shorten product development cycle, effectively improve production efficiency, and promote the transformation of production mode to customization, decentralization and service.

Intelligent manufacturing covers intelligent products characterized by intelligent interconnection, intelligent production with intelligent factory as the carrier, intelligent management with information physical system as the key, and intelligent service characterized by real-time online. The intelligent manufacturing system covers all aspects of product design, production planning, production execution, after-sales service and so on. The physical system is the foundation of intelligent manufacturing, and intelligent factory is the key to intelligent manufacturing.

Application of intelligent manufacturing technology in industrial automation

Intelligent manufacturing technology has been widely used in the field of industry, especially in the field of industrial automation. In the field of industrial automation, intelligent manufacturing technology can show the following advantages.

The man-machine operation technology

The intelligent manufacturing technology has achieved high precision, high quality and high efficiency in the field of industrial automation. It is impossible to realize the traditional automation control. Even manual operation cannot be realized. For example, the precision of some metal products in the space and aviation field is not very strict, the artificial manufacturing cannot be achieved, and it can be used. The computer is connected with the NC machining equipment, and the numerical control machine is transmitted by the program instruction compiled by the computer, which is automatically produced by the numerical control equipment to ensure the precision requirements of the product.

The automatic design

The intelligent manufacturing system has powerful functions of reasoning, prediction, judgment and so on. The system can automatically design industrial products according to the digital signal or program code identified by the outside world. In actual production, the product R & D personnel only need to transfer the key parameters and program code of a product to the intelligent manufacturing system. The intelligent manufacturing system can be used. To automatically design and display the product model on the computer, and let the R & D personnel choose the best solution according to the requirements. For example, many enterprises adopt CAD, Pro.E, UG and other automatic design software, which can be connected with the intelligent manufacturing system, and the designed product model is more accurate[5].

The virtual production

The intelligent manufacturing system can use the computer to simulate the production. That is, the system simulates the production process of the industrial product according to the requirement of the enterprise, combining the signal processing technology, the animation technology, the intelligent reasoning and the simulation technology. Even if the problem of the design product is found, it is easy to improve the production process and can be used at the same time. The raw material and production time required for production are controlled.

The development trend of China's intelligent manufacturing

Suit one's measures to local conditions

To promote intelligent manufacturing to promote the transformation and upgrading of the manufacturing industry, to accelerate the development of advanced manufacturing industry, and to become an effective way to make a move towards a powerful manufacturing country, it is necessary to adapt to local conditions, adapt to current conditions, adapt to the conditions of enterprises, to be practical, and to achieve success.

In order to promote industry

By speeding up the "machine replacement", "machine generation" and "equipment replacement", the pace of "equipment replacement" has formed a strong demand for intelligent manufacturing, thus promoting the rapid development of the intelligent manufacturing equipment industry, and further promoting the development of the equipment manufacturing industry and the whole manufacturing industry.

Rammed Foundation

(1) To speed up the development of the intelligent equipment industry, to develop the intelligent equipment, key parts and key software.

(2) To improve the management foundation of the enterprise and combine with the lean production.

Integrated traction

Take the system integration supplier as the lead, connect the components, the single machine, the hardware and software to the user, and provide the customer with the automatic line, the intelligent manufacturing system, the digital workshop and the intelligent factory's overall solution.

Grasp the time sequence

Grasp the development rhythm, grasp the development time sequence between the intelligent manufacturing application and the intelligent device industry development, coordinate the development rhythm between the two, and win the time and development space for the intelligent manufacturing equipment and products of the domestic brand.

Master the core

Master the core technology of intelligent devices, intelligent manufacturing equipment and systems, and build an independently controllable domestic intelligent manufacturing equipment industry.

Demonstration and guidance

Intelligent manufacturing is the development direction of the manufacturing industry, and the manufacturing industry has a process to realize intelligent manufacturing, and the manufacturing industry has different levels of intelligent manufacturing. Therefore, the industry must be guided by the demonstration industry, through the ladder, sector development, and in the industry.

Construction of cluster

Intelligent manufacturing involves many fields, enterprises, disciplines and technology, with complex supporting relations and systematicness. Only building industrial ecology conducive to the development of intelligent manufacturing, forming an organism of intelligent manufacturing cluster, can be effectively promoted, and the development of interregional development should avoid the development of homogenization.

Setting up an innovation center

In our country, there is a huge gap between the basic technology of intelligent manufacturing, the development of system software, the ability of original innovation and the developed countries of foreign countries. It is necessary to draw lessons from the experience of the construction of innovative network in the United States and promote the construction of the innovation system of intelligent manufacturing in our country by innovating the concept of development and driving the development of our country, and setting up the government as the dominant, the enterprise as the main body, the research organization and the society. In different areas, different areas, different areas, the establishment of intelligent manufacturing basic technology innovation research center, intelligent manufacturing industry software innovation research center, intelligent system integration research center, intelligent equipment innovation research center, and other research centers related to the above major categories. The level research talents participate in the creative activities of intelligent manufacturing, push the leading role of the enterprise in the development of manufacturing technology, and guide the social capital to the intelligent research and development process of the manufacturing industry.

Words

industrial	*adj*. 工业的，产业的；*n*. 工业；工人
revolution	*n*. 革命；彻底改变；旋转；运行，公转
intelligent	*adj*. 聪明的；理解力强的；有智力的 *n*. 智能；聪明的；理解力强的；智慧的
continuous	*adj*. 连续的；延伸的；连绵
advanced	*adj*. 先进的；高深的；年老的；晚期的 *v*. 前进；增加；上涨
methods	*n*. 方法；方法；方法论；教学法；分类法
effectively	*adv*. 有效地；实际上，事实上
physical	*adj*. 自然（界）的；身体的；物质的 *n*. 身体检查，体格检查
sensing	*n*. 感觉；传感；感知；遥感
aviation	*n*. 航空；飞行术；飞机制造业；飞机
automatic	*adj*. 自动的；无意识的；必然发生的 *n*. 自动化机器或设备；自动手枪
virtual	*adj*. 实质上的，事实上的；虚拟的
enterprise	*n*. 企（事）业单位；事业
material	*n*. 材料，原料；*adj*. 物质的
trend	*n*. 走向；趋向；时尚，时髦；*vi*. 倾向；趋势
promote	*vt*. 促进，推进；提升
foundation	*n*. 地基；基础；基金（会）

sequence	*n.* 数列，序列；顺序；连续；片断；插曲 *vt.* 使按顺序排列，安排顺序
rhythm	*n.* 节奏，韵律；节律，规律；节拍；调和
innovation	*n.* 改革，创新；新观念；新发明

Notes

[1] It can be considered that intelligent manufacturing will be the leading industry leading the third industrial revolution.
【注释】此句为 that 从句，其中，It 指代主语“intelligent manufacturing”。

[2] Intelligent manufacturing is not a new thing, it is the continuous popularization of information technology and gradually developed in the 80s of 20 century. Advanced manufacturing technology and computer technology are widely used in modern manufacturing.
【注释】此句为省略句，用 it 指代主语“intelligent manufacturing”。

[3] Intelligent manufacturing not only adopts new manufacturing technology and equipment, but also infiltrates the rapidly developing information and communication technology…
【注释】“…not only…, but also…”表示“不仅……而且……”

[4] Intelligent manufacturing is a new type of manufacturing system, which combines advanced automation technology, sensing technology, control technology, digital manufacturing technology…
【注释】此句为 which 从句，其中，which 指代主语“intelligent manufacturing”。

[5] For example, many enterprises adopt CAD, Pro.E, UG and other automatic design software, which can be connected with the intelligent manufacturing system, and the designed product model is more accurate.
【注释】此句为 which 从句，其中，which 指代主语“many enterprises”。

Part B Reading Materials

Industry 4.0 and Smart Manufacturing

Industry 4.0 is a technological transformation of an industry and a change of industry. The intelligent manufacturing proposed by Industry 4.0 is aimed at the product life cycle and realizes the information manufacturing under the ubiquitous sensing condition. Intelligent manufacturing technology is based on modern sensing technology, network technology, automation technology and artificial intelligence. Through the sensing, human-computer interaction, decision-making, execution and feedback, the product design process, manufacturing process and enterprise management and service intelligence are realized. It is the deep integration and integration of information technology and manufacturing technology.

The biggest difference between intelligence and automation is the amount of knowledge. Intelligent manufacturing is based on science, not just on experience. Scientific knowledge is the foundation of intelligence. Therefore, intelligent manufacturing involves both material and non-material processing processes, not only has a sound and responsive material supply chain, but also requires a stable and strong knowledge supply chain and industry-university-research alliance, providing high-quality talents and industry continuously. The required innovations, the development of the new products with high are added value, and the continuous transformation and upgrading of the industry.

Smart manufacturing is a sustainable manufacturing model that leverages the enormous potential of computer modeling and information and communication technologies to optimize product design and manufacturing processes, dramatically reducing material and energy consumption and waste generation. Recycle and reduce emissions and protect the environment.

The smart factory based on the concept of Industry 4.0 will consist of a physical system and a virtual information system called the Cyber Physics Production System (CPPS), which is a blueprint for tomorrow's manufacturing industry.

Part C Exercises

Translate the following paragraphs into Chinese

Manufacturing industry is not only the cornerstone of a country's economic development, but also the basis of strengthening the national competitiveness. It is also the carrier of multi-professional high and new technology, which reflects the overall level of national high and new technology. With the wide application of artificial intelligence technology such as expert system, neural network and genetic algorithm in manufacturing system and its various links, it is possible to obtain, express, transfer, store and reason manufacturing information and knowledge. A new production mode of manufacturing intelligence has emerged. Intelligent design, intelligent machining, robot, intelligent control, intelligent process planning, intelligent scheduling, intelligent measurement and so on are the main aspects of intelligent manufacturing. Intelligent manufacturing technology is an organic combination of traditional manufacturing technology, computer technology, network technology, automation technology and artificial intelligence technology. The gradual maturity of artificial intelligence and its successful application in manufacturing industry is the key technology guarantee to realize intelligence in manufacturing industry.

Lesson 29 Advanced Processing Equipment

教学目的和要求

本文为介绍先进加工设备的英文专业文献，主要围绕先进加工设备相关技术的定义与专业术语的英文表达进行介绍。通过本文的学习，读者可以了解有关先进加工设备中常见专业词汇的英文表达和专业术语的英文名称，有助于全面了解先进加工设备所涉及的专业知识；要求掌握文中所涉及的先进加工设备相关术语、相关学科的英文表达，并能对文后所附的练习进行理解和翻译。

重点和难点

（1）重点掌握先进加工设备相关专业术语及其英文表达。

（2）了解先进加工设备的应用领域及其发展趋势。

Part A Text

CNC Machine

Since the advent of the first CNC machine in the 1950s, CNC machine and CNC technology have experienced more than half a century of development. Nowadays, CNC machining has spread to various manufacturing fields, which not only improves the quality and efficiency of product processing, but also shortens the production cycle. It has improved working conditions and has had a profound impact on the product structure and production methods of manufacturing companies[1]. As an efficient and precise digital cutting technology, CNC machining has become the main means of mechanical machining of complex structural parts of aircraft. More than 50% of the machining work of aircraft structural parts is completed by CNC machining. With the continuous development of the aviation industry[2], aircraft performance continues to increase, aircraft structural parts are becoming larger and more complex, and more stringent requirements are imposed on the corresponding CNC machining equipment and CNC machining technology.

The intelligentization of CNC machine tools is the index-controlled machine tool that can obtain the processing related information such as strain, vibration and thermal deformation generated during the machining process, to automatically compensate and optimize the machining performance of the machine tool to improve the machining precision, surface quality and machining of the CNC machining center. At present, many machine tools are gradually integrated with adaptive control systems such as Artis, which can realize overload protection of tools and machine tools not only through machine spindle monitoring (tool balance detection, conflict detection, bearing detection, etc.) and tool monitoring (breakage detection, wear detection, etc.)[3]. Moreover, real-time speed regulation can be performed through data acquisition during processing and automatic judgment of the system, thereby achieving efficient processing of stable loads.

The main factors that restrict the quality and efficiency of CNC machining are on the one hand, the hardware conditions of the machine tool, and more importantly, the related enabling technologies that support high-efficiency CNC machining, such as tool technology and tooling technology.

1) Development of Tool Technology

Tool technology is one of the key technologies of CNC machining, and it is also a technical bottleneck that limits the rate of CNC machining centers for difficult-to-machine materials. With the advancement of tool technology, the tool material and tool structure are continuously improved, and the types of tools are more and more. How to choose the tool and cutting parameters reasonably is the core of improving the efficiency of CNC machining.

2) Tooling technology development

At present, the installation method of large-scale aviation structural parts in China is relatively simple: aluminum alloy structural parts mainly adopt reserved process ear pieces, and use screw compression or vacuum adsorption difficult-to-machine materials such as titanium alloy are mainly pressed by press plates honeycomb core materials is mainly bonded by double-sided tape. CNC fixtures have used a large number of automatic fixtures with pneumatic, hydraulic and control systems. The use of CNC multi-point automatic adjustment, vacuum adsorption or mechanical chuck flexible clamp can realize the flexibility, rapid positioning and clamping of large-sized structural parts with different shapes on the machine tool. It has become the development direction of CNC tooling design and manufacture, and it is to improve the numerical control. Another key technology is for processing efficiency.

CNC Machining Center

The machining center was developed from a CNC milling machine. The biggest difference with the CNC milling machine is that the machining center has the ability to automatically exchange machining tools. By installing different tools on the tool magazine, the machining tool on the spindle can be changed in one clamping by the automatic tool changer to realize various machining.

CNC machining center is a high-efficiency automatic machine tool composed of mechanical equipment and numerical control system for processing complex parts. CNC machining center is one of the most widely used and widely used CNC machine tools in the world. Its comprehensive processing ability is strong, the work piece can complete more processing content after one clamping, and the processing precision is high. The batch work piece with medium processing difficulty is 5~10 times more efficient than ordinary equipment, especially it can be completed. The processing that cannot be completed by ordinary equipment is more suitable for single-piece process or small-to-medium batch multi-species production with complex shapes and high precision requirements. It combines the functions of milling, boring, drilling, tapping and cutting threads on one piece of equipment, giving it a variety of processes. Machining centers are classified according to the spatial position of the spindle machining: horizontal and vertical machining centers. It is classified by process use: boring and milling machining center, composite machining center[4]. Special classification by function: single workbench, double workbench and multi-workbench machining center. Machining centers for single-axis, dual-axis, three-axis and interchangeable headstocks.

Flexible Manufacturing System

In recent years, the rapid development of computer technology, microelectronics technology and mechanical equipment manufacturing technology has led to major changes in modern manufacturing. Although the traditional automated manufacturing technology has high production efficiency, it cannot meet the demand of short-cycle, multi-variety and small-volume product manufacturing in today's market, that is, the demand for flexible production. Therefore, flexible manufacturing technology has gradually become a modern manufacturing industry.

The flexible manufacturing system is an automated manufacturing system that connects processing equipment with a transmission device and an automatic loading and unloading device under the unified control and management of a computer system, and is suitable for high-efficiency processing of small and medium-sized batches and multi-variety parts.

1) Flexible manufacturing unit

This unit is a machining unit consisting of one or more CNC machine tools or machining centers. The unit can be automatically replaced with tools according to production needs to adapt to the processing of different work pieces. The unit is generally suitable for processing parts with small batch size, complicated shape, simple process and long processing time. The unit's equipment is highly flexible, but personnel and processing flexibility is low.

2) Flexible automatic production line

The line is a line of equipment between high-volume, single-variety and non- flexible automated production lines and small to medium-sized and multi-variety flexible manufacturing systems. Generally, the production line can connect multiple machine tools with the material transportation system, and the flexibility of the material transmission system is lower than that of the flexible manufacturing system, but the production efficiency is high.

3) Flexible manufacturing system

The system refers to a production system consisting of a CNC machine tool or a machining center, plus a material transportation system. It is mainly controlled by a computer and can continuously process a variety of work pieces. The system is mainly suitable for the management and production of small and medium-sized parts, complex shapes and multi-variety parts.

Ultra-precision Machining

Ultra-precision machining technology plays an irreplaceable role in national defense construction and national economic development. It is an important supporting technology for modern high-tech warfare and an important guarantee for the development of modern basic science and technology. Ultra-precision machining technology plays an increasingly important role in many high-tech fields such as aerospace, precision instruments, military industry, optical and electronic communication, new energy, etc., and several major issues in the National Medium- and Long-Term Science and Technology Development Plan. Special items such as "Manned Space Flight and Lunar Exploration Project" and "High-end CNC Machine Tools and Basic Manufacturing Equipment" are directly related. It is the technical basis for the implementation of several major national science and technology projects. In a sense, ultra-precision machining technology is one of the important indicators to measure a country's scientific and technological strength. Ultra-precision machine tools are the key carrier for ultra-precision machining, which directly determines the accuracy, efficiency and reliability of part machining. Over the years, all countries in the world have attached great importance to the development of ultra-precision machine tools and their processing technology. They have made great progress in both military and civilian fields and reached a high level.

At present, the ultra-precision processing technology is mainly in the United States, the United Kingdom, Japan, etc. For this reason, they have set up specialized research institutions to develop special research plans, such as the United States developed research for ultra-precision machine tools POMA plan in the 1980s. In the development of ultra-precision machine tools, Germany, Switzerland, the Netherlands and South Korea are also advanced in addition to the United States, the United Kingdom and Japan. European countries such as Russia, Ukraine, the Czech Republic and France also have good technical skills. In the early 1980s, China began research on ultra-precision machining technology and machine tools. It started late. In the past 30 years, through the efforts of the government and related research units, great progress has been made, and some aspects have reached the world advanced level. However, compared with foreign developed countries, there is still a big gap. The development trend of ultra-precision machine tool technology is generally developing in the direction of extreme, intelligent, green and service. It should be pointed out that extreme, intelligent, green, and serviced are not completely independent, but have mutual integration.

Words

impact	*n.* 碰撞，冲击，撞击；影响；冲击力
	vt. 挤入，压紧；撞击；对……产生影响
	vi. 冲撞，冲击；产生影响
precise	*adj.* 清晰的；精确的；正规的；精密

corresponding	*adj.* 相当的，对应的；一致的
	v. 相符合；类似；相配
obtain	*vt.* 获得，得到；流行；买到，达到
	vi. 通行，通用；流行；存在
deformation	*n.* 变形；形变；失真；应变
monitoring	*n.* 监视；控制；监测；追踪
judgment	*n.* 判断；审判；判决；判断力
hardware	*n.* 硬件
flexible	*adj.* 灵活的；易弯曲的；柔韧的
equipment	*n.* 设备，装备；器材，配件
multiple	*adj.* 多重的；多个的；复杂的；多功能的
	n. 倍数；并联
defense	*vt.* 防御；防守；辩护；抗辩
modern	*adj.* 现代；现代的；新式的
optical	*adj.* 视觉的，视力的；眼睛的；光学的
extreme	*adj.* 极端的，过激的；极限的；*n.* 极端；困境

Notes

[1] It has improved working conditions and has had a profound impact on the product structure and production methods of manufacturing companies.
【注释】此句为省略句，其中，It 指代“CNC machine”。

[2] With the continuous development of the aviation industry …
【注释】此句为长句式，用 With 引导的这个从句作为后面句子的主语。

[3] At present, many machine tools are gradually integrated with adaptive control systems such as Artis, which can realize overload protection of tools and machine tools…
【注释】此句为 Which 引导的从句，其中，which 指代“adaptive control systems”，作为从句的主语。

[4] It is classified by process use: boring and milling machining center, composite machining center.
【注释】此句为省略句，其中，It 指代“CNC machining center”。

Part B Reading Materials

Intelligent Manufacturing and Advanced CNC Technology

Under the influence of intelligent manufacturing environment, advanced numerical control technology is moving towards open intelligence, while high-grade CNC machine tools are used as material carriers for numerical control technology. Typical functions include intelligent measurement, real-time compensation, processing optimization, tool management, remote monitoring and Diagnosis, etc. As an independent manufacturing unit of integrated manufacturing system, high-end CNC machine tools, driven by common standard interfaces and network interconnection, artificial intelligence technology and advanced manufacturing technology, also promote the development of integrated intelligent manufacturing systems The integrated intelligent manufacturing system, such as the rapid development of intelligent workshops and smart factories, naturally marks the successful completion of the intelligent manufacturing demonstration project.

The advanced numerical control technology represents the core competitiveness of high-end CNC machine tools. It mainly has two typical characteristics—open and intelligent, and also includes important features such as function compounding, greening, integration, and digitization. The open main index control system is equipped with a standardized basic platform, providing standard interfaces and internetwork, allowing developers to intervene in different software and hardware modules, with modularity, portability, scalability and interchangeability intelligent main index Control machine tools and CNC systems have intelligent functions such as intelligent processing, intelligent monitoring, intelligent maintenance, intelligent management, and intelligent decision making. High-end CNC machine tools are important integrated application equipment for numerical control technology. Relevant advanced numerical control technologies mainly involve intelligent programming, multi-axis linkage, high-speed high-precision control, machine tool error compensation, process parameter self-optimization, adaptive control, online diagnosis and remote maintenance, and intelligence, production management, machine tool network group control management, etc.

Part C　Exercises

Translate the following paragraphs into Chinese

Driven by intelligent manufacturing, advanced numerical control technology has developed rapidly. In addition to the above-mentioned high-speed high-precision linkage control technology, machine tool error compensation technology and intelligent control technology, intelligent programming technology, intelligent numerical control management technology, intelligent people Machine interfaces and the like are also widely concerned. CNC machine tools are the material carrier of numerical control technology. The advancement of advanced numerical control technology naturally promotes the research and development of intelligent high-end CNC machine tools, and has become the research focus of high-end numerical control equipment in the field of intelligent manufacturing. In China, according to the classification of intelligent manufacturing equipment industry clusters with high-end CNC machine tools as the core, equipment R&D and production enterprises are mainly distributed in the Bohai Rim region, the Yangtze River Delta region, the Pearl River Delta and the Northwest region. For enterprises, the most typical numerical control enterprises include Huazhong CNC, Guangzhou CNC, Shenyang Machine Tool, Pushing Ningjiang, Qinchuan Machine Tool, Dalian Machine Tool, Qizhong CNC.

Unit Seven Scientific Paper and Equipment Instruction

Lesson 30 How to Write a Scientific Paper

教学目的和要求

本文介绍科技论文的写作内容要求、通用的格式和结构，以及它们在论文中的作用和意义。通过课文的讲解，使读者了解科技论文的写作特点和规范，并能根据本文的学习撰写规范的科技论文。

重点和难点

（1）重点学习科技论文的结构和写作规范。
（2）重点掌握科技论文摘要的写作。
（3）了解科技论文的语言表达特点，学习复合句的表达。

Part A Text

A scientific paper is a written and published report describing original research results. That short definition must be qualified, however, by noting that a scientific paper must be written in a certain way and it must be published in a certain way.

Title

In preparing a title for a paper, the author would do well to remember one salient fact: That title will be read by thousands of people. Perhaps few people, if any, will read the entire paper, but many people will read the title, either in the original journal or in one of the secondary (abstracting and indexing) services. Therefore, all words in the title should be chosen with great care, and their association with another must be carefully managed.

What is a good title? We can define it with the fewest words that adequately describe the contents of the paper.

The title of a paper is a label. It is not a sentence. Because it is not a sentence, with the usual subject, verb, object arrangement, it is really simpler than a sentence (or, at least, usually shorter), but the order of the words becomes even more important.

The meaning and order of the words in the title are of importance to the potential reader who read the title in the journal table of contents. But these considerations are equally important to all potential users of the literature, including those (probably a majority) who become aware of the paper via secondary sources. Thus, the title should be useful as a label accompanying the paper itself, and it also should be in a form suitable for the machine-indexing systems used by the Engineering Index, Science Citation Index, and others. Most of the indexing and abstracting services are geared to *key word* systems. Therefore, it is fundamentally important that the author provide the right *keys* to the paper when labeling it.That is, the terms in the title should be limited to those words that highlight the significant content of the paper in terms that are both understandable and retrievable.

Abstract

An abstract is a concise and precise summary of the paper. The role of the abstract is not to evaluate or explain, but rather to describe the paper (dissertation). The abstract should include a brief but precise statement of the problem or issue, a description of the research method and design, the major findings and their significance, and the principal conclusion[1]. The abstract should contain the most important words referring to method and contend of the paper: These facilitate access to the abstract by computer research, and enable readers to identify the basic content of a document quickly and accurately, to determine its relevance to their interests, and thus to decide whether they need to read the document in its entirety[2].

An abstract should be written in complete sentences, rather than in phrases and expressions. Generally, an abstract for a short paper is limited to a maximum of 200～250 words. The abstract should be designed to define clearly what is dealt with in the paper. Many people will read the abstract, either in the original journal or in The Engineering Index, Science Citation Index or one of the other secondary publications.

The abstract should never give any information or conclusion that is not stated in the paper. References to the literature must not be cited in the abstract (except in rare instances, such as modification of a previously published method).Because the abstract is not a part of the paper, it is neither numbered nor counted as a page.

Introduction

Now that we have the preliminaries out of the way, we come to the paper itself. I should mention that some experienced writers prepare their title and abstract after the paper is written, even though by placement these elements come first. You should, however, have in mind (if not on paper) a provisional title and an outline of the paper that you propose to write. You should also consider the level of the audience you are writing for, so that you will have a basis for determining which terms and procedures need definition or description and which do not.

The first section of the text proper should, of course, be the introduction. The purpose of the introduction should be to supply sufficient background information and the design idea to allow the reader to properly understand and evaluate the results of the present study without needing to refer to previous publication on the topic. The introduction should also provide the rationale for the present study. Above all, you should state briefly and clearly your purpose in writing the paper. Choose references carefully to provide the most important background information.

Suggested rules for a good introduction are as follows：(1) It should present first, with all possible clarity, the nature and scope of the problem investigated. (2) It should review the pertinent literature to orient the reader. (3) It should state the method of the investigation. If deemed necessary, the reasons for the choice of a particular method should be stated. (4) It should state the principal results of the investigation. (5) It should state the principal conclusion(s) suggested by the results.

Materials and Methods

Now, in Materials and Methods, you must give the full details. Most of this section should be written in the past tense. The main purpose of the Materials and Methods section is to describe the experimental design and then provide enough detail that a competent worker can repeat the experiments. Many (probably most) readers of your paper will skip this section, because they already know (from the introduction) the general methods you used and they probably have no interest in the experimental detail. However, careful writing of this section is critically important because the cornerstone of the scientific method requires that your results, to be of scientific merit, must be reproducible. And, for the results to be adjudged reproducible, you must provide the basis for repetition of the experiments by others. That experiments that are unlikely to be reproduced are beside the point.The potential for producing the same or similar results must exist, or your paper does not represent good science.

When your paper is subjected to peer review, a good reviewer will read the Materials and Methods carefully. If there is serious doubt that your experiments could, he repeated, the reviewer will recommend rejection of your manuscript no matter how awe-inspiring your results.

For materials, include the exact technique specifications and qualities and source and method of preparation. Generally, it is necessary to list pertinent chemical and physical properties of specimens (or reagents) used.

For method, the usual order of presentation is chronological. Obviously, however, related methods should be described together, a straight chronological order cannot always be followed. If your method is new (unpublished) you must provide all of the needed detail. However, if a method has been previously published in a standard journal, only the literature reference should be given.

Results

So now we come to the core of the paper, the data. This part of the paper is called the Results section. There are usually two ingredients of the Results section. First, you should give some kind of overall description of the experiments, providing the *big picture*, without, however, repeating the experimental details previously provided in Materials and Methods. Second, you should present the data.

Of course, it isn't quite easy. How do you present the data? A simple transfer of data from laboratory notebook to manuscript will hardly do. Most important, in the manuscript you should present representative data rather than endlessly repetitive data.

The Results need to be clearly and simply stated, because it is the Results that comprise the new knowledge that you are contributing to the world. The earlier parts of the paper (Introduction, Materials and Methods) are designed to tell why and how you got the Results. The later part of the paper (Discussion) is designed to tell what they mean. Obviously, therefore, the whole paper must stand or fall on the basis of the Results. Thus, the Results must be presented with clarity.

Discussion

The Discussion is harder to define than the other sections. Thus, it is usually the hardest section to write. And, whether you know it or not, many papers are rejected by journal editors because of a faulty Discussion, even though the data of the paper might be both valid and interesting. Even more likely, the true meaning of the data may be completely obscured by the interpretation presented in the Discussion, again resulting in rejection.

What are the essential features of a good Discussion? I believe the main components will be provided if the following injunctions are heeded.

(1) Try to present the principles, relationships, and generalizations shown by the Results. And bear in mind, in a good Discussion, you discuss—you do not recapitulate the Results.

(2) Point out any exceptions or any lack of correlation and define unsettled points. Never take the high-risk alternative of trying to cover up or fudge data that do not quite fit.

(3) Show how your results and interpretations agree (or contrast) with previously published work.

(4) Don't be shy, discuss the theoretical implications of your work, as well as any possible practical applications.

(5) State your conclusions as clearly as possible.

(6) Summarize your evidence for each conclusion.

In showing the relationships among observed facts, you do not need to reach cosmic conclusions. Seldom will you be able to illuminate the whole truth. More often, the best you can do is shine a spotlight on one area of the truth. Your one area of truth can be buttressed by your data. If you extrapolate to a bigger picture than that shown by your data, you may appear foolish to the point that even your data-supported conclusions are cast into doubt.

When you describe the meaning of your little bit of truth, do it simply. The simplest statements evoke the most wisdom. Verbose language and fancy technical words are used to convey shallow thought.

Words

salient	*adj.* 突出的，显著的，卓越的，优质的，明显的
adequately	*adj.* 充分地，适当地
citation	*n.* 引用
table of contents	目录
retrievable	*adj.* 可获取的，可取回的，可重新获得的，可恢复的，可挽救的
dissertation	*n.* （学位）论文，专题，论述，学术演讲
relevance	*n.* 关联，关系，适用，中肯
preliminary	*adj.* 预备的，初步的
provisional	*adj.* 暂定的，假定的，暂时的，临时的
rationale	*n.* 基本原理，理论基础，原理的阐述
above all	尤其是，最重要的是，首先是
pertinent	*adj.* 有关的，相干的，中肯的
cornerstone	*n.* 基石；基础；（建筑）隅石
manuscript	*n.* 手稿，原稿
awe-inspiring	令人敬畏的，令人鼓舞的
specification	*n.* 详述；[*pl.*]规格，说明书；规范；明细表
specimen	*n.* 试样，样品；标本
reagent	*n.* 反应物，反应力，试剂
chronological	*adj.* 按时间顺序排列的；按年代顺序排列的
ingredient	*n.* 成分，要素，因素，原料
obscure	*adj.* 模糊的，含糊的，晦涩的，暗的，朦胧的；*v.* 使……黑暗，使不明显
injunction	*n.* 命令，指令
heed	*v.* & *n.* 注意，留心
recapitulate	*v.* 扼要重述，概括，重现，再演

unsettled	*adj.* 不稳定的，不安定的，未解决的，混乱的
correlation	*n.* 关联，相关性，相互关系
cover up	包裹，隐藏，掩盖
fudge	*n.* 捏造，梦话，胡话，空话；*vi.* 蒙混，逃避责任；*vt.* 粗制滥造，捏造；推诿；*int.* 胡说八道
implication	*n.* 牵连，受牵累；暗示，隐含；意义，本质
cosmic	*a.* 宇宙的，全世界的，广大无边的
illuminate	*v.* 照明，照亮，阐明，说明，着凉，使光辉灿烂，以灯火装饰（街道等）；*vi.* 照亮
spotlight	*n.* 聚光灯，点光源，公众注意中心
buttress	*n.* 支持物，支柱；*v.* 支持，加强，扶住
extrapolate	*n.* 推断，外推，外插
evoke	*v.* 唤起，引起，博得，移送
verbose	*adj.* 冗长的，累赘的，喋喋不休的
fancy	*adj.* 奇特的，美妙的，漂亮的；*n.* 想象力，嗜好，爱好

Notes

[1] The abstract should include a brief but precise statement of the problem or issue, a description of the research method and design，the major findings and their significance，and the principal conclusion.

【译文】摘要应该包括简洁而精确的关于问题或论点的陈述、研究方法及设计思路的描述、主要研究结果和它们的意义，以及主要的结论。

【注释】该句中的 a brief but precise statement of the problem or issue、a description of the research method and design、the major findings and their significance 及 the principal conclusion 均为 include 的宾语。

[2] The abstract should contain the most important words referring to method and contend of the paper：these facilitate access to the abstract by computer research, and enable readers to identify the basic content of a document quickly and accurately, to determine its relevance to their interests, and thus to decide whether they need to read the document in its entirety.

【注释】"these facilitate access to…, and enable readers to…entirety"为并列句，facilitate 与 enable 为并列的两个谓语。该句中的 to identify…, to determine…及 to decide…均作为 reader 的宾语补足语。

Part B Reading Materials

Writing Abstracts

Abstracts are short, informative writings that serve as screening tools or previews for research papers, conference presentations, and other communications. Abstracts' focus (summary vs. results) and format (headings vs. no headings) vary across contexts. Writers in all contexts give their texts similar functions such as *reviewing literature*, *identifying a problem* or *describing an approach*. These functions are called rhetorical moves.

Identifying Rhetorical moves

Below are five rhetorical moves that can appear in abstracts, questions these moves may answer, and sample sentences from an abstract in technology. A move's length can vary (from a phrase to sentences) and some moves may be omitted depending on the abstract's audience and purpose.

Move 1: Introducing background or problem

What is currently known? What is the gap in knowledge?

Examples: Children undergoing long-term hospital care face problems of isolation from their familiar home and school environments.

This isolation has an impact on the emotional wellbeing of the child.

Move 2: Presenting current research with justification and/or purpose

What is this study's aim? How does it fill the gap in knowledge?

Examples: In this paper we report on research that explores the design of technologies that mitigate some of the negative aspects of separation, while respecting the sensitivities of the hospital, school and home contexts.

This paper reports on the field trial of the technology.

Move 3: Describing methodology

How was the study conducted? Was the data quantitative, qualitative, or both?

Examples: We conducted design workshops with parents, teachers and hospital staff…

In response we designed a novel technology that combined an ambient presence with photo-sharing to connect hospitalised children with schools and families.

Move 4: Reporting results

What were the outcomes? What was discovered?

Example: We ... found that there was a strong desire for mediated connection, but also a significant need to protect privacy and avoid disruption.

Move 5: Interpreting results

How are the results interpreted? How has this study contributed to the field?
Example: The research provides new insights into how technology can support connectedness and provides a foundation for contributing to the wellbeing of children and young people in sensitive settings.

Complete Abstract

Children undergoing long-term hospital care face problems of isolation from their familiar home and school environments. This isolation has an impact on the emotional wellbeing of the child. In this paper we report on research that explores the design of technologies that mitigate some of the negative aspects of separation, while respecting the sensitivities of the hospital, school and home contexts. We conducted design workshops with parents, teachers and hospital staff and found that there was a strong desire for mediated connection, but also a significant need t protect privacy and avoid disruption. In response we designed a novel technology that combined an ambient presence with photo-sharing to connect hospitalised children with schools and families. This paper reports on the field trial of the technology.

Part C Exercises

1. Translate the following Abstracts into English.

（1）例题一。

一种平面四杆机构优化设计方法的研究

摘要：文中介绍了一种依据平面机构的基本组成原理，构造出平面连杆机构的数学模型并通过 VC 编程的方式快速地得到其运动学特性的方法。该方法有利于产品开发过程中的参数优化设计。
关键词：连杆机构；参数优化；平面四杆机构

（2）例题二。

挖掘机器人铲斗连杆机构优化研究

摘要：在计算理论挖掘力与分析铲斗挖掘时的整机挖掘性能时，铲斗机构的总传动力臂成为制约其性能发挥的关键因素。文中利用遗传算法并结合实际作业时铲斗的工作频谱，对铲斗机构的结构尺寸和铰点位置进行优化，建立了优化数学模型。在保证空间有效作业范围的同时，可以较大程度地增大传动力臂，提高有效挖掘力，并结合某型挖掘机进行改造，验证了优化的有效性，提高了实际挖掘力。
关键词：挖掘机器人；传动力臂；遗传算法；工作频谱

2. Translate the following Abstract into Chinese.

Summary on the review and development trend of computer aided fixture designing technology
Abstract: The computer aided fixture design(CAFD)has already becoming an important part of

CAD / CAM integration technology developed science the 70 years of the 20th century up to now. This paper proceed with 4 researching aspects(installation planning, clamping planning, fixture configuration design, fixture performance evaluation)included in the CADF technique, a review has been carried out on the development results mainly within the near 10 years of the development of CAFD techniques at home and abroad, and an analysis was carried out on the trend of future development of CAFD.

Key words: computer aided fixture design, installation planning, clamping planning, fixture configuration design, fixture performance evaluation.

Lesson 31 Translation of Equipment Instruction

教学目的和要求

本文为高梯度磁选机英文材料的翻译实例，由于篇幅有限，只节选了前三节的内容。这部分内容比较有代表性，反映了机械产品设备的说明书通用的格式和内容，包括概述、注意事项和安全准则。关于设备的具体结构和操作原理以及控制部分，属于供货商和采购设备的厂家版权所有，未经许可不能公开。因此，在本文中省略这一部分。

通过本文的讲解，使读者了解设备说明书的格式、规范和主要涵盖的内容，了解说明书中常见的安全标识和注意事项的英文表达，为以后在工作中翻译设备说明书打下基础。

重点和难点

（1）重点学习设备说明书的通用格式和写作规范。

（2）重点掌握各种安全标识和注意事项的表达特点。

（3）准确把握说明书的语言特点，例如，警示语的表达宜言简意赅且具有鲜明的强调作用。

Part A Text

High-gradient magnetic separator HGS 150 40 S15 2 is shown in Fig. 31.1.

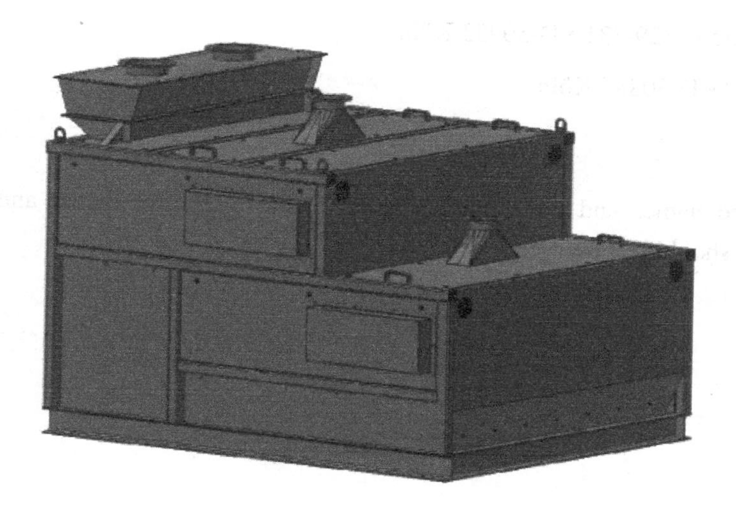

Fig.31.1 HGS 150 40 S15 2

Before starting any work, read the instructions, note the components fitted and observe the supplier documentation attached[1]!

These instructions contain important information for safe and efficient work with the machine. These instructions form part of the machine and must always be stored to hand and in a clearly legible condition. Anyone operating or working on the machine must be able to access them.

Before starting work always ensure that anyone operating or working on the machine has carefully read and understood these instructions. Observance of and compliance with all the safety notices and guidelines contained in these instructions are essential for safe and efficient work with the machine.

Observe and comply with applicable local accident prevention requirements and general safety rules.

The figures in these instructions have been provided for illustrative purposes and for general understanding. They may differ from the current model. Requirements cannot be derived from the figures in these instructions.

Translation of the Original

Printed in Germany

Manufacturer and customer service

Should you have any technical questions or need spare parts, please contact:

Steinert Elektromagnetbau GmbH

Widdersdorfer Straße 329-331 • D-50933 Köln

Postfach 45 11 60 • D-50886 Köln

Brand names

Third-party brand names and trade marks belong to the respective owner and the associated trademark rights should be observed.

1. Overview

1.1 Information about the instructions

These instructions contain important information about working with the machine. Compliance with all the safety notices and guidelines provided is essential for safe work[2].

➲ Keep right next to the machine

➲ Make accessible to staff at all times

➲ Keep in a clean and legible condition

➲ Read carefully before starting any work

➲ Observe accident prevention requirements and general safety rules.

1.2 Limitation of liability

All details and information provided in these instructions have been produced taking account of the applicable standards and specifications, state-of-the-art technology, and the knowledge and experience we have gained over the years.

The manufacturer does not accept any liability for damage caused by:

- Failure to observe the instructions
- Incorrect application
- Use of untrained staff
- Unauthorized alterations
- Technical modifications
- Use of spare parts which have not been approved

Furthermore, the duties agreed on in the delivery contract, the General Terms and Conditions of Business, the manufacturer's conditions of delivery, and the legal requirements applicable at the time the contract is concluded all apply.

We reserve the right to technical modifications to improve performance characteristics and for further development.

1.3 Copyright protection

Treat these instructions as confidential. They are intended solely for the staff who work with the machine. The instructions must not be given to third parties without the written approval of the manufacturer.

The details, texts, drawings, illustrations, and other figures they contain are protected by copyright and are subject to commercial property rights. Any form of misuse is punishable by law.

Reproductions of any kind, in part or in whole, and the use and/or disclosure of the contents are not permitted without the written consent of the manufacturer. In the event of contraventions, we will claim compensation for damages. We reserve the right to assert further claims.

1.4 Warranty conditions

The warranty conditions are provided as a separate document with the sales documents.

1.5 Customer service

Incorrect spare parts!

Damage to property

➲ Use genuine spare parts from the manufacturer.

Our customer service is happy to provide technical information.Information about the responsible contacts is available at any time by phone, fax, e-mail or on our website.

Steinert Elektromagnetbau GmbH

Widdersdorfer Straße 329-331 • D-50933 Köln

Postfach 45 11 60 • D-50886 Köln

Head office: +49 (0) 221 / 49 84 – 0

Service and spare parts: +49 (0) 221 / 49 84 – 100 (and – 177)

Fax: +49 (0) 221 / 49 84 – 219

E-mail: sales@steinert.de

http://www.steinertglobal.com/grp/de/

Our staff is always interested to learn about new information and hear about experiences with our products which could be beneficial for further improvements.

2. Symbols

2.1 Warnings

Warnings are marked by symbols in these instructions. The information is introduced by signal words indicating the degree of risk.

Type and source of danger

Possible consequences of failure to comply with warning

➲ Actions to avoid the danger.

Classification

Warns of an imminent danger resulting in death or serious injury unless avoided.

Warns of a potentially dangerous situation resulting in death or serious injury unless avoided.

▲ CAUTION	Warns of a potentially dangerous situation resulting in moderate or minor injury unless avoided.

2.2 Notes — damage to property and the environment

NOTE	Warns of a potentially dangerous situation, which will cause damage to property and the environment unless avoided.

•

•

2.3 Information

Information!

Highlights recommendations and information relating to efficient and smooth operation.

2.4 Other symbols

Guidelines

Structure of guidelines:

➲ Instruction for an action

⇨ Result if needed

Lists

Structure of lists without numbers:

■ List level 1

■ List level 2

3. Safety

This section provides an overview of all important safety aspects for optimum staff protection and for safe and smooth operation.

Failure to comply with the guidelines and safety notices provided in these instructions may result in serious dangers.

Electric voltage!

Risk of fatal electric shock.

- Only allow installation and commissioning to be undertaken by qualified electricians or under the guidance of the manufacturer[3].
- Only allow work on the electrical system to be carried out by electricians.
- Only operate if you have not undertaken unauthorized alterations and the system is in perfect technical condition.
- If the insulation is damaged, switch off the voltage supply immediately and arrange for repairs.
- Never bridge or disable fuses.
- Only replace fuses with equivalent ones.

Static magnetic field!

Risk of serious injuries or death for people with metallic and/or active medical implants.

➲ Ensure that people with active medical implants do not enter the direct or indirect hazard zone.

➲ Ensure that people with metallic implants do not enter the direct or indirect hazard zone.

DANGER

Dynamic and static load!

Risk of death.

➲ Have the suitability of the foundations and installation site checked by a structural engineer.

➲ For self-supporting constructions, note the dynamic loads produced by the machine's vibrations and movements.

➲ Only use aids with sufficient dimensions.

DANGER

Risk of ignition from highly flammable substances!

Serious injury or death due to fire or explosion.

Within the hazard zone and in the direct vicinity of the system:

- Do not smoke.
- No naked flames or any kind of sources of ignition.
- Provide fire extinguishers.
- Report suspect substances, liquids or gases to the responsible officer immediately.

In the event of fire:

- Stop work straight away.
- Immediately leave the hazard zone until you are told it is safe to return.

WARNING

Risk of ignition!

Risk of injury due to fire or explosion.

- Do not smoke.
- No naked flames or any kind of sources of ignition.
- Keep hazard zone free of dust.
- Stop work immediately if a lot of dust is produced.
- Wait until the dust has settled.

⇨ Then,

- remove layer of dust.

⚠ WARNING

Risk of material falling or being flung out!

Risk of injury.

➲ Do not enter the hazard zones when in automatic mode.

➲ Wear personal protective equipment if entering the hazard zones in manual mode.

⚠ WARNING

Noise!

Damage to hearing.

➲ Wear hearing protection.

➲ Only enter the noise zone when necessary.

➲ Use appropriate means of communication.

⚠ WARNING

Illegible safety signs!

Risk of injury.

➲ Replace illegible safety signs immediately.

⚠ CAUTION

Hot surface!

Burns.

➲ Avoid contact.

➲ Wear appropriate personal protective equipment.

➲ Allow hot surfaces to cool to ambient temperature.

⚠ CAUTION

Hot operating materials!

Burns.

➲ Avoid contact.

➲ Wear appropriate personal protective equipment.

⚠ CAUTION

Sharp edges and corners!

Risk of injury.

➲ Wear appropriate personal protective equipment.

⚠ CAUTION

Tripping hazards!

Risk of injury.

➲ Keep work area clean.

➲ Remove objects that are no longer needed.

➲ Identify tripping hazards as specified.

3.1 Responsibility of the operator

The machine is used in a commercial environment. The operator therefore has statutory health and safety duties.

➲ Comply with the valid safety, accident prevention, and environmental protection specifications for the area of use.

➲ Retrofit the required safety equipment.

➲ Obtain information about prevailing health and safety requirements.

➲ In a risk assessment, ascertain hazards resulting from the special work conditions at the machine's site of operation.

➲ Define measures to prevent risks in the form of instructions for operating the machine.

➲ Whenever the machine is being used, check that the operating instructions provided comply with the latest rules and regulations and adapt if necessary.

➲ Clearly regulate and define responsibilities for installation, operation, maintenance, and cleaning.

➲ Ensure that all staff working on the machine have read and understood the operating instructions.

➲ Train staff at regular intervals and inform them of hazards.

➲ Provide staff with the protective equipment needed.

- ➲ Ensure that the machine is always in perfect technical condition.
- ➲ Ensure that the maintenance intervals described in the operating instructions are observed.
- ➲ Regularly check that all safety equipment is fully functional and complete.

3.2 Safety equipment

3.2.1 Speed sensor

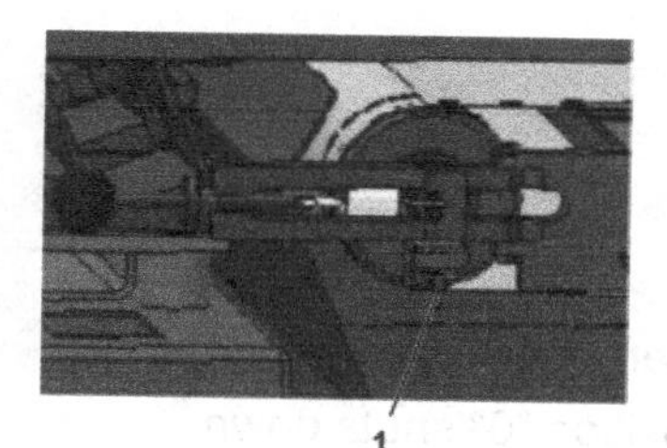

An optional sensor is placed at the separation module to monitor the belt path.

The speed is evaluated via the controller.

For the model with a sensor, refer to the supplier documentation in the Appendix.

3.2.2 Emergency stop button

Unauthorized switching back on!

Serious injury or death.

Before switching back on:

- ➲ Ensure that only authorized staff can switch the machine back on.
- ➲ Ensure that there is not anyone in the hazard zone.

Pressing an emergency stop button triggers the emergency stop.

⚠ DANGER

Unauthorized switching back on!

Serious injury or death.

Before starting any work:

➲ Switch off.

➲ Lock to prevent switching back on.

➲ Ensure that only authorized staff can switch the machine back on.

➲ Ensure that there is not anyone in the hazard zone.

Once all work is complete:

➲ Fit all covers.

- The main switch is not an emergency stop switch. Turning the main switch to position "0" shuts down the power supply.

Main switch positions

0 = OFF

I = ON

Lock main switch to prevent switching back on

Depending on the design of the main switch, it may be locked in position "0" with a padlock to prevent it being switched back on.

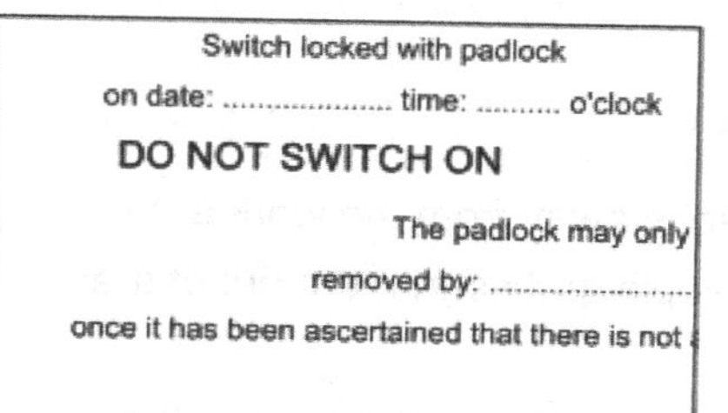

Lock main switch as follows to prevent switching back on:

- ➲ Switch off main switch.
- ➲ Lock main switch with a padlock.
- ➲ Fit sign on switch where it is clearly visible.
- ➲ The employee stated on the sign holds the key.

Shut off
date: .... time: .... o'clock
DO NOT SWITCH ON
May only be switched on
by: ....
once it has been ascertained
that there is not anyone
in the hazard zone.

If the switch cannot be locked with a padlock:

- ➲ Switch off main switch.
- ➲ Fit sign on switch where it is clearly visible.

3.3 Staff requirements

3.3.1 Qualification

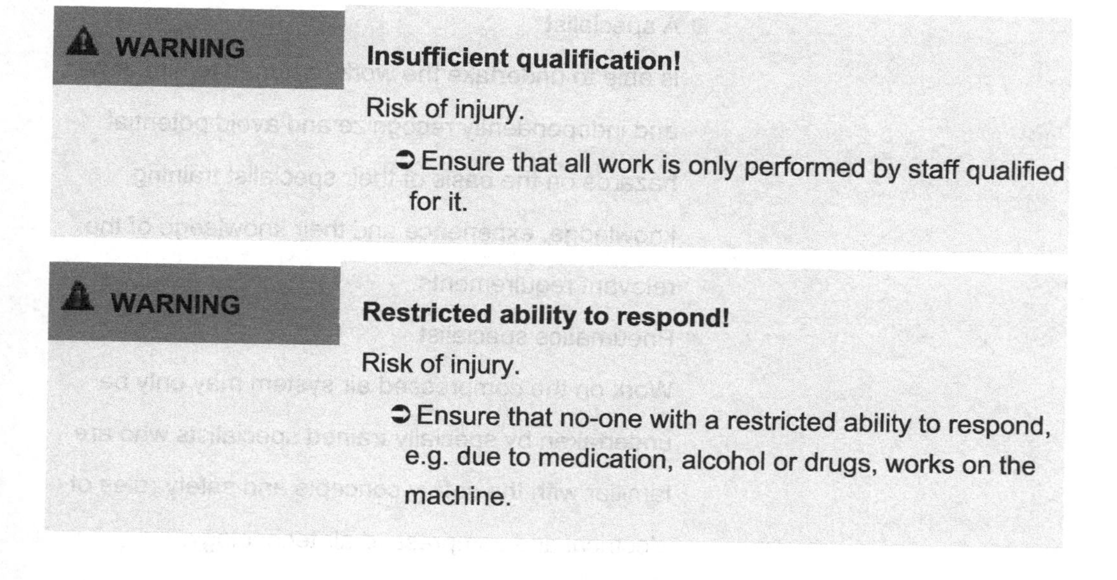

WARNING **Insufficient qualification!**

Risk of injury.

➲ Ensure that all work is only performed by staff qualified for it.

WARNING **Restricted ability to respond!**

Risk of injury.

➲ Ensure that no-one with a restricted ability to respond, e.g. due to medication, alcohol or drugs, works on the machine.

WARNING **Unauthorized persons!**

Risk of injury.

➲ Keep unauthorized persons away from the work area.

➲ If in any doubt, talk to people and send them out of the work area.

➲ Stop work if there are any unauthorized people in the work area.

Improper actions may result in significant injury or damage to property.

➲ When selecting staff, note the requirements relating to age and occupation applicable at the site of use.

The following qualifications are used in the instructions for various areas of work:

■ An instructed person

has been instructed by the operator in an instruction session about the work assigned to him or her and possible risks resulting from improper actions.

■ A specialist

is able to undertake the work assigned to him or her and independently recognize and avoid potential hazards on the basis of their specialist training, knowledge, experience and their knowledge of the relevant requirements.

■ Pneumatics specialist

Work on the compressed air system may only be undertaken by specially trained specialists who are familiar with the safety concepts and safety rules of electrical and compressed air technology.

- An electrician

  is able to undertake work on the electrical systems and independently recognize and avoid potential hazards on the basis of their specialist training, knowledge, experience and their knowledge of the relevant standards and requirements.

  The electrician is trained for the particular site of operation where he or she works and is familiar with the relevant standards and requirements.

- Unauthorized persons

  are people who do not satisfy the above requirements (instructed person, specialist).

3.3.2 Instruction

➲ Regularly instruct staff in how to work safely with the machine.

➲ Log instruction sessions provided

Date	Name	Type of instruction	Instruction provided by	Signature

3.4 Personal protective equipment

➲ When working on the machine, use the protective equipment needed for the work.

➲ Observe information about personal protective equipment provided in the work area.

➲ Do not wear rings, necklaces, and other jewelry.

Wear protective clothing

Use foot protection

Wear hand protection

Wear safety goggles

Wear ear defenders

Wear a hard hat

Wear lightweight breathing apparatus

➲ When undertaking special work, also wear the following personal protective equipment. During special work

Using a safety harness

➲ Safety harnesses may be used by specially trained personnel only.

3.5 Limit values for magnetic field strengths to protect people

- The max. permissible limit value for people with active medical implants is 1 mT (the equivalent of 10 gauss). Protection from magnetic fields
- This value is not exceeded at a distance of 35 m from the strongest magnet.
- For people without implants and limited exposure, the maximum permissible value is 2 T (20,000 gauss).
- The permissible average value for continuous exposure (approx. 8 h) is 212 mT.
- The max. values measured at a distance of approx. 1 m from a magnet with the highest field strengths and a power of 70 kW were less than 100 mT.

3.6 What to do in hazard and accident situations

Preventive measures

- Keep first-aid equipment (first aid box, blankets etc.) and fire extinguishers to hand.
- Familiarize staff with equipment for reporting accidents, first aid, and rescue.
- Keep access routes clear for emergency services vehicles.

What to do in hazard and accident situations

- In hazard situations, stop machine movements as quickly as possible and switch off the power supply.

In a hazard situation, proceed as follows:

- Initiate an emergency stop immediately by pressing the nearest emergency stop button.
- Rescue people from the hazard zone.
- Administer first-aid.
- Switch off main switch and lock to prevent switching on again.
- If necessary, call a doctor and the fire service.
- Inform responsible people at the site of operation.
- Keep access routes clear for emergency services vehicles.

After rescue:

- If required by the severity of the incident, inform the authorities responsible.

3.7 Environmental protection

NOTE	**Lubricants!** Damage to the environment. ➲ Arrange for authorized specialist firm(s) to handle disposal. . ***If substances harmful for the environment are released into the environment:*** ➲ Clear up contamination. ➲ Inform the municipal authorities responsible.
NOTE	**Electronic scrap, electronic components, lead components!** Damage to the environment. ➲ Arrange for authorized specialist firm(s) to handle disposal.

3.7.1 Lubricants

➲ Ensure that lubricants and greases are not released into the environment.

➲ Arrange for specialist disposal firm to dispose of lubricants and greases.

3.8 Signage

The following symbols and safety signs are used in the work area. They relate to the direct surroundings in which they are fitted.

3.8.1 Type plate

The type plate is on the basic chassis and contains the following details:

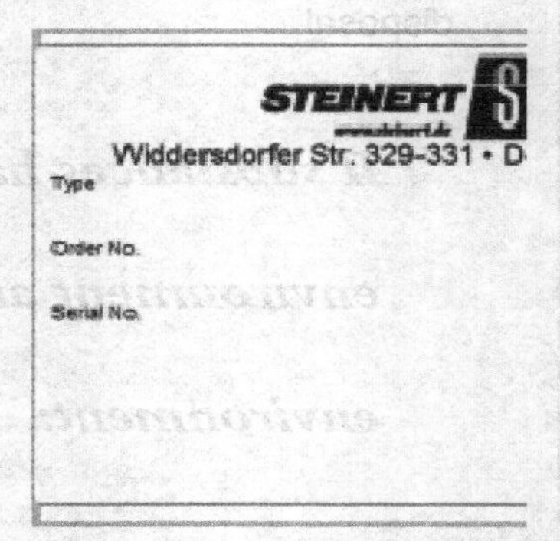

Manufacturer with address

- Type
- Year of manufacture
- Commission number
- Serial number
- Connection voltage
- Weight

3.8.2 Symbols and warning signs

⚠ WARNING	**Illegible safety signs!** Risk of injury. ➲ Replace illegible safety signs immediately.

Observe directions for use

Point of attachment

Electric voltage

Hot surface

Warning of falling objects

Ban for people with pacemakers

Warning of magnetic field

Center of gravity

Warning of hazard spot

Words

high-gradient magnetic	高梯度磁选机
instruction	*n*. 授课；教诲；传授的或获得的知识；[计算机科学]指令
limitation	*n*. 限制；局限；极限；起限制作用的规则（或事实、条件）
state-of-the-art	*adj*. 使用最先进技术的；体现最高水平的
copyright protection	版权保护
warranty conditions	保修条件
customer service	客户服务
electric voltage	电压
static magnetic field	静磁场
dynamic and static load	动静态载荷
ignition	*n*.（汽油引擎的）发火装置；着火，燃烧；点火，点燃
appendix	*n*. 附录；阑尾；附加物
lubricant	*n*. 润滑剂，润滑油；能减少摩擦的东西
signage	*n*. 标记；标识系统

Notes

[1] Before starting any work, read the instructions, note the components fitted and observe the supplier documentation attached!

【译文】请仔细阅读说明书，注意部件的安装并遵守供应商的说明要求！

[2] These instructions contain important information about working with the machine. Compliance with all the safety notices and guidelines provided is essential for safe work.
【译文】这份说明书包含控制器工作的重要信息。为保证安全工作，请遵守说明书中所有的指南和安全注意事项。

[3] Only allow installation and commissioning to be undertaken by qualified electricians or under the guidance of the manufacturer.
【译文】安装和调试必须在专业电工或制造商的指导下进行。

Part B Reading Materials

Proper use

The machine is intended for separating magnetizable and paramagnetic contaminants from bulk materials or obtaining resources from it.

Only feed bulk materials to the machine that comply with the material properties specified/approved by the manufacturer for this order.

- For the composition of the bulk materials, refer to the order confirmation.

1. Site of operation

Roofed stand, hall

2. Reasonably foreseeable incorrect use

Any form of use above and beyond proper use and/or any other kind of machine use may result in dangerous situations.

Never use the machine:

- without protective clothing
- for storing material
- to supply materials of the wrong particle size
- to supply pasty, liquid or highly flammable substances

There is no ground for claims of any kind for damage resulting from improper use.

The operator assumes sole liability for all damage resulting from incorrect use.

Part C　Exercises

Translate the following table contents.

Technical specifications

High-gradient magnetic separator	HGS 150 40 S15 2
Metering channel* Jöst	Type FDL 1500/-50×675 with support
Channel width	1500 mm
Type of drive	Metering oscillator of magnetic drive
Drive	4×JD 50
Output	4×0.7 kW
Frequency	50 Hz
Connection voltage	230 V
Separation module	1 and 2
Transport bandwidth	approx. 1745 mm
Belt speed (electronically adjustable)	0.8 to max. 2.0 m/s
Drive * SEW	For each separation module
Speed	161 rpm
Output	P= 0.37 kW; n= 161 rpm; model M1A;
Frequency	50 Hz
Connection voltage	220～240 / 380～415 V
Protection rating	IP54
Magnetic pulley	BRP 150 S15
Diameter	150 mm
Weight HGS assembly	approx. 1750 kg
Weight, FDL	185 kg
Weight, separation module	approx. 265 kg
Weight, magnetic pulley	approx. 160 kg